AF477158

ENERGY RESOURCES
CONVENTIONAL & NON-CONVENTIONAL

SECOND EDITION

ENERGY RESOURCES
CONVENTIONAL & NON-CONVENTIONAL
SECOND EDITION

Dr. M. V. R. Koteswara Rao

Guest Faculty, Centre for Environment, IST,
Jawaharlal Nehru Technological University, Hyderabad - 500 028.

BS Publications

A unit of **BSP Books Pvt., Ltd.**

4-4-309/316, Giriraj Lane, Sultan Bazar,
Hyderabad - 500 095
Phone : 040 - 23445605, 23445688

Published by :

BS Publications

A unit of **BSP Books Pvt., Ltd.**

4-4-309/316, Giriraj Lane, Sultan Bazar,
Hyderabad - 500 095
Phone : 040 - 23445605, 23445688
e-mail : info@bspbooks.net

ISBN : 978-93-52300-10-5 (HB)

Preface to the Second Edition

The first edition of the textbook, "Energy Resources, Conventional and Non-Conventional" has good response from students and teachers pursuing education in the fields of chemical sciences, environmental technology and management. The need for incorporating information on new energy-delivering materials such as biofuels, high energy material-hydrogen, direct energy conversion techniques, fuel cells usage in technology programmes and finally energy audit, has thus arisen.

In this edition, a good account on the latest technologies coming up in the fields of energy systems and management has been added at appropriate places in several chapters. Indeed these are futuristic topics that assume importance in the high technology.

In this connection I would place on record the help rendered by my colleague, Prof. P.V. Krishna Rao, in writing a chapter on "Direct Energy Conversion" (Chapter 15) for the benefit of students and scientific community.

I am also grateful to Prof. Y. Anjaneyulu, and Prof. A.V.S. Prabhakara Rao, for constant encouragement and discussions.

- Author

Preface to the First Edition

The inter-disciplinary subject of ENERGY AND ENVIRONMENT covers both environmental resources and the generation of energy in nature. In order to understand the subject, thorough knowledge of the environmental setting, ecology and the dynamics of energy transfer is necessary. The basics of the subject are common for engineering and postgraduate environmental science &technology students. I drafted this book from the course material which I dealt with for the postgraduate students over several years.

The subject is discussed in 16 chapters. The first 3 chapters deal with environment topics, the influence of environment while generating different forms of energy, definitions of work and power, as well as the laws of thermodynamics. Later chapters briefly cover fossil fuels, solar, hydro, ocean, wind, geothermal, biomass, electrochemical and nuclear types of energy. The present state of knowledge on the use of high energy materials such as Hydrogen, the Fuel cells, the Solid Polymer Electrolysis (SPE) and the batteries has also been discussed. The energy scenario pertinent to India has also been dealt with and conclusions drawn.

The importance of conservation of energy as well as introduction of energy-efficient gadgets in household and industry has been spelt out. Latest literature from research papers, reports, internet data and books has been compiled in this book, thus making it an up-to-date manual.

It is hoped that this text book on Energy and Environment would broadly serve the students, teachers and the research scholars community.

Dr. M.V.R. Koteswara Rao
Hyderabad.

Contents

CHAPTER 5

SOLAR ENERGY

CHAPTER 6

HYDRO ENERGY

CHAPTER 7

OCEAN ENERGY

CHAPTER 8

WIND ENERGY

CHAPTER 9

GEOTHERMAL ENERGY

CHAPTER 10

BIOMASS

CHAPTER 11

BATTERIES AND ELECTRIC POWER

CHAPTER 12

NUCLEAR ENERGY

CHAPTER 13

CONVERSION OF MATTER INTO MORE USEFUL FORMS

CHAPTER 14

STORAGE OF ENERGY

CHAPTER 15

DIRECT ENERGY CONVERSION

CHAPTER 16

ENERGY SCENERIO - INDIA

CHAPTER 17

ENERGY AUDIT

1

INTRODUCTION

Energy is the capacity to do work. The natural resources of energy are: *the solar radiation, wind,, rivers and ocean resources, geothermal fluids, atomic minerals and biomass.* Environment is a setting of these energy resources that keep on changing according to the withdrawal of energy. All living bodies in the universe need both energy and environment for their life. Also both of these elements are inter-related. In order to derive energy a congenial environment is essential and the resources of energy create a sustained environment through various transformations such as photosynthesis, mineral deposition, fossil fuels generation, rains and other means.

Historically, man in ancient times, that is during the stone age, was aware of the heat energy as the outcome of fires from the woody trees. Also they used to derive some power through the work of horse, oxen, elephant and other domestic animals. Even the muscle power of man has been spent for cutting trees and laying the roofs of house shelters.

Slowly the realisation that the environment is a provider of energy has set in his mind. The reference of environment as a setting for energy systems has been made in ancient Hindu mythology as well as in Bible. It was believed in ancient times that the earth was resting on a tortoise and all the elements required for life, namely the "space, land, air, rain, fire etc." (the *panchabhutams*) were governed by God from the fulcrum. During later years, Sun was considered the chief source of energy in the universe and adored as God. Even in Holy Bible a reference was made to the Sun. Milton in *"Paradise Lost"* longed for light to be seen by him. Rain, a source of energy and an useful element for life was referred many a time in the *vedas* and epics of Hindus. The mountains *"Himalayas"* and *Vindhyas"* in India were recognised as godly since they preserve the energy resources. In epics, Pravara realised that the snowy peaks of the Himalayas were the embodiment of vigorous energy. It was stated in Mahabharat that Agasthya maharshi brought the river Cauvery from heaven down to earth. Similar report that the turbulent river Ganga was brought from the head of God Shiva onto earth by Bhagiratha maharshi was made in epics. That the energy resources were assembled and harnessed by people of valour was clearly described in the epic Mahabharata giving the example of building the capital city Indraprastha. The people during this period knew the art of utilising the energy resources to their advantage. In fact the houses and towns were designed at that time in accordance to the principles scientifically laid down in *"Vasthu"*, the science of architecture of ancients. Following this, it appears that the submerged capital city of Lord Krishna, namely the Dwaraka, was built approximately during the epic times. Furthermore, maintenance of ecology and biodiversity in Nature was a ritual in Hindu life.

SUSTAINABLE ENVIRONMENT

The three components of Nature that are to be understood at this stage, to study the inter-relationship between environment and energy are: *the ecosysem, the micro- environment and the natural resources of energy*. The structure of the ecosystem comprises of the living species and the physical world surrounding them.

The living species are animals, plants and human beings. The physical world represents their dwellings in rural, urban, forest, mountain, desert, farmland, marine and polar areas. The living bodies depend mostly on these surroundings for their sustenance. The natural resources, besides generating the requisite energy, provide food too. The structure of the ecosystem extends in the atmosphere, hydrosphere and the lithosphere and consists of both biotic and abiotic objects. The micro environment represents the habitat's physical surroundings and the places of his activity only. This setting is liable to be polluted due to the activity and hence all care has to be taken to sustain the quality of this environment with minimum effort. As man uses various types of available energies and consume the food resources around, a depletion of energy at a particular location takes place. It is necessary to sustain the environment at that location by adopting suitable methods. The inter-relationship between environment and energy can be easily visualised following the diagram shown in Fig. 1.1.

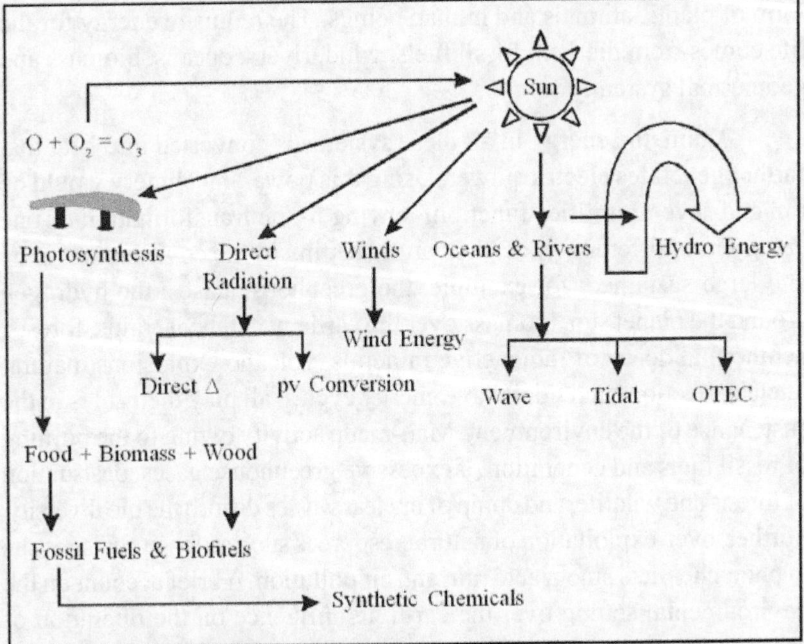

Fig. 1.1 *Inter-relationship between energy & environment*

The primary source of energy in the universe is the Sun. The receptor is the Earth. Both the sun and earth are satellites moving independently in space separated by a distance of 93,000,000 mi and maintaining their gravitational forces of attraction and repulsion. The diameter of the Sun is 863,000 mi and that of the earth is 77910 mi. The sun is a hot body and maintains a temperature of around 6000 K. Around the sun all materials exist in the gaseous phase only. The glow of the satellite called the ray is transmitted on all sides and carries heat energy. The rays pass through different media before they reach the surface of earth. It takes about 8 minutes for a ray to reach the earth. The solar energy is thus transmitted on the earth. The solar flux is 1300 watts per sq.meter per minute.

The earth maintains reasonable temperatures at many places (between 40-100 °F) which are congenial for living bodies. Also, the gaseous atmosphere around the earth balances the climate required for the life. So the earth is unique amongst the satellites to nurture life in the form of plants, animals and human beings. The requisite energy for the life comes from the Sun, fossil fuels, wind, rivers, oceans, biomass and geothermal system.

Again, the energy in all these systems is converted into heat that further generates electric power. Using this power, machinery would be run and several utilities function. Owing to the transformation of one energy into different types, the natural setting in the environment is by and large sustained; for example, the greenhouse gases, the hydrogen around the planet sun, biomass over the earth, watersheds, rains, forests, geothermal decay of radioactive minerals, volcano explosions, natural nuclear reactions and tidal/wave energy cycles, all play their roles in the sustenance of the environment. Man-made activity owing to the burning of fossil fuels and generation of excessive greenhouse gases, destruction of forests and wildlife, and dump of nuclear wastes disturb the bio diversity. Further, over-exploitation of natural resources such as fossil fuels results in photochemical smog, acid rain and air pollution. A brief account on the environmental setting over the earth, its influence on the liberation of energy has been outlined below.

THE EARTH

Planet Earth is small and unique amongst the stellar. It has a surface that continuously gets renewed. The surface temperatures are congenial for life to set in. There are many water resources on earth that are required for the living bodies. The life on earth is mostly based on carbon and nitrogen and these are provided by the greenhouse gases around the earth and the plant life.

Structure of Earth

The diameter of earth is 27 miles shorter than the diametric length of equatorial axis, if viewed along the poles. Therefore the central portion of the earth is slightly squeezed in shape. The picture displaying the structure of earth is shown in Fig. 1.2. Compared to the large radius of the earth, the outermost layer namely the crust, is rather thin; it is about 3 miles thick under the oceans and 21 miles thick under the continents. Down below the crust lies the "mantle" that is about 1860 miles thick and this constitutes of rocks, sediment and silicate bodies. Even the water is entrapped in the layers of the mantle. Then we have the "core". It is divided into 'outer core' and 'inner core'.

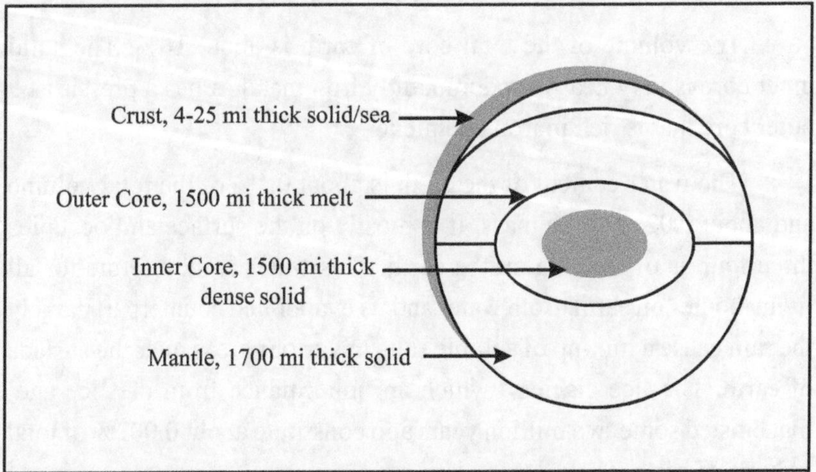

Fig. 1.2 Structure of Earth

The core is the heaviest portion of the earth, containing chiefly iron, nickel, mercury (in small measure only) and other metals. The outer core has free-flowing liquid metal held under high pressure, whereas the inner core is a solid mass. As the liquid swirls, earth's magnetic field is generated. The temperature of the core is around 5000 °C.

The Earth Materials

The crust of the earth is about 0.8% of total volume of earth. The continental crust is granite rock on the surface and underlying sedimentary and metamorphic rocks below the surface. All these are hard. The oceanic crust is mostly 'basalt' rock made of silicate rich in iron and magnesium.

The mantle is a major portion of the earth and it constitutes about 83% of earth's volume. It is composed of dense silicate minerals rich in iron and other metals such as copper, barium, zinc, lead, aluminium, beryllium, silver and gold; amongst the non-metals sulphur is prominent. Owing to volcanic eruptions that take place in the earth, there exists a mix-up of the minerals brought by magma from the outer core. Even radioactive minerals are thus entering into the crust of the earth.

The volume of the total core of earth is about 16%. The solid inner core is very heavy predominantly iron metal. It has a pressurised outer core that is rich in iron and nickel.

The water content of the earth is about 0.1% of the total volume and about 0.02% of the mass. It is mostly on the surface and occupies three fourths of the area on the earth. This water is the lifeline for all living bodies on earth. Both water and its evaporated counterpart namely the rain cause a mix-up of soluble salts in various areas over the surface of earth. The 'ice glaciers' which are inherittance from the 'ice age' that existed some two million years ago constitute about 0.002% of total volume of earth.

Earth's Atmosphere

The various gases and vapours that surround the surface of the earth is termed 'atmosphere'. This has been sub-divided into different layers taking into consideration the composition of gases, the temperature gradient and the density of air. The immediate layer that surrounds the earth upto 17 km in the atmosphere is 'troposphere'. This region consists of 75% of the earth's air. Above this layer is situated the 'stratosphere', the second layer starting from 17-48 km. Subsequently the 'mesoshpere' between 50- 80 km and the 'thermosphere' at 90 km onwards exist.

The original gases that surrounded the earth were carbon dioxide and nitrogen. That the atmosphere around earth was created from the original algae plants got a proof from the presence of these gases. Also, earth's atmosphere has argon, Ar40 that generated from the decay of radioactive K40, but not the Ar38 or Ar36 which would have originated from solar systems. In this respect the atmosphere around the earth is unique in the sense that these gases maintain the congenial conditions for all living bodies. The gas oxygen was a latter entry in the atmosphere and after millions of years it reached a level of 21% today. The troposphere subsequently developed a thin layer of ozone that was regulating the ultraviolet radiation emanating from the sun towards the earth. The volcano mud and the biomass brought out these environmental transformations over the years. The ocean currents and the winds formed from solar radiations over the earth distributed the heat in the continents.

The fertile lands that are necessary for the plant growth and foods, the mineral wealth for day-to-day progress of mankind have also been identified and exploited. The volcano eruptions and underwater explosions on earth are responsible for the changes in earth's structure.

Volcanoes and Environmental Change

The earth's *oceans and atmosphere* is created from the volcanic activity that took place over millions of years and these two natural elements are constantly sustaining the life and giving energy. In a way, volcanoes destroy the habitation and side by side create the congenial environment also.

People were attracted by the mud and flanks of the volcanic soils and were deriving the Nature's fruits throughout the world. Today, it appears that about 80% of the earth's surface over the land and seas originated from volcanoes only.

Volcanoes are formed by eruption from its own energy, namely the heat given out by the magma of the earth. The magma is formed on account of the high temperature prevailing in the earth's upper mantle, which permits the rocks to melt down and flow as 'blobs' of magma through the fractures. A few of these blobs recombine and form 'reservoirs' or 'holding tanks'. The underlying pressure in the earth pushes the hot magma to erupt through the weak tracks of mantle and the whole mass gushes out in the form of 'lava'. The magma that reaches the surface and flows is the lava and when it cools down at the surface it forms solidified igneous rocks.

The characteristic of magma is that it contains dissolved gases that build up pressure and at the same time is instrumental in the generation of rocks of 'pumice' type. The gases also determine the explosive nature of the volcanic eruption depending on the viscosity of the lava. Molten magma contains metals such as calcium, sodium, potassium, titanium and magnesium. Lava contain gases such as carbon dioxide, sulphur oxides.

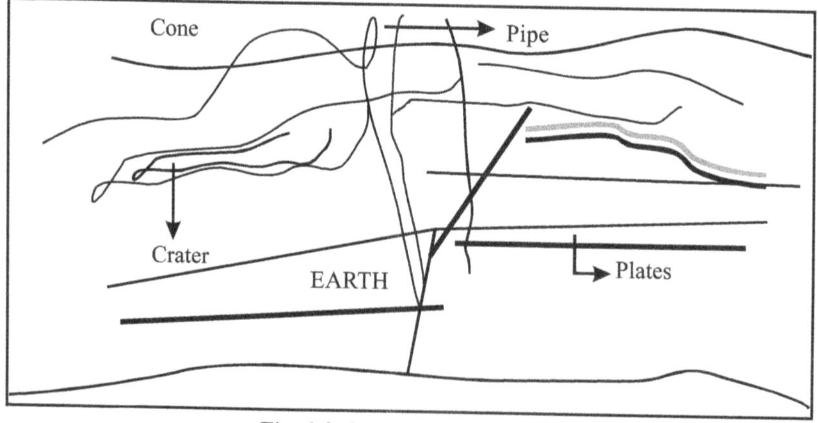

Fig. 1.3 Structure of a Volcano

The essential features of a volcano are shown in the line drawing (Fig.1.3). In the upper mantle hot magma is generated and accumulated in a reservoir. A volcano vent in the earth allows the magma to enter. A pipe adjoining the vent makes the passage way for the molten magma to rush out to the surface of the earth. On the floor of earth the cool lava assumes usually a cone shape with intermittent rocks, ashes and spreads of muds. Also the lava enters some craters of earth where they deposit the ash and mud.

Based on the shape and the materials of build-up, volcanoes are classified into the following:

- composite volcanoes
- shield volcanoes
- cinder cones
- spatter cones
- complex volcanoes

The magma of a volcano erupts in different ways in the regions of world; even under the ocean volcano eruptions do take place and these are known as 'submarine eruptions'. The hot mixture of gas, ash and the rocks that travel along the slope of a volcanic eruption is termed 'pyroclastic flows'. This extends to several miles at times causing devastation. The 'Lahars' are mudflows that are mixtures of dusts and water and they also flow with speed; whatever object comes in the way is destroyed by the hot speedy mass. Volcanic blasts are those which are irregular and assymmetric bursts in the cone. Even 'debris avalanches' or 'landslides' are also resultant of this type of blast.

Geysers, fumaroles (also known as solfatoras) and hot springs are formed by the percolation of surface water through the crevices into the rocks down to the mantle of earth where the hot magma still persists or might have just solidified but remained hot. The water that is in contact suddenly becomes hot steam and rises back to the earth's surface through the fissures in the form of springs. The geysers of "old faithful" at Yellowstone National Park pose an excellent scenery.

Mankind has to put up with the currently active volcanoes, 500 in number. Most of them are situated in the Pacific region and some of them exist under the ocean. The volcano provides a good energy resource namely, 'geothermal energy' which is clean and inexhaustible. The mud and ash that is thrown out during the eruptions turn out to be fertile soil. Also the mineral wealth of the earth grows in those regions of the world, eg. Japan . The supply of elemental sulphur for the global countries is met on account of the geo-formation of this element. The liberation of carbon dioxide and other gases add on to the greenhouse gases and warm the atmosphere to a certain extent.

Global Warming

The solar radiation (mostly in ultraviolet and visible range of 180-800nm) incident on the earth causes several changes in the energy pattern of the atmosphere. A portion of the radiation is reflected back into the space (about 6%). The clouds that remain in between reflect about 17% of the incoming radiation. About 8% of incoming solar radiation is back-scattered by the air. Out of the remaining, 19% is absorbed by the ozone, water vapour and particle dust; 4% by clouds located in the lower troposphere and 47% directly incident on the surface of the earth. These short wave radiations heat up the earth significantly.

However an equilibrium temperature of about 300 Kelvins are set up automatically in the radiation heating of earth. In order to constantly maintain this pattern of heating , the earth has to re-emit some energy into the space through long wave length infrared radiation. From the out-going infrared radiation 15% is directly radiated by the cloud-free earth surface, 60% radiated back into the space by the atmosphere and the clouds. An overall 7% for sensible heat flux and 24% latent heat flux is retained near the surface of earth that is gradually re-radiated. This was the 'thermal balance' existing for over centuries upto the beginning of 1900 (shown in Fig. 1.4).

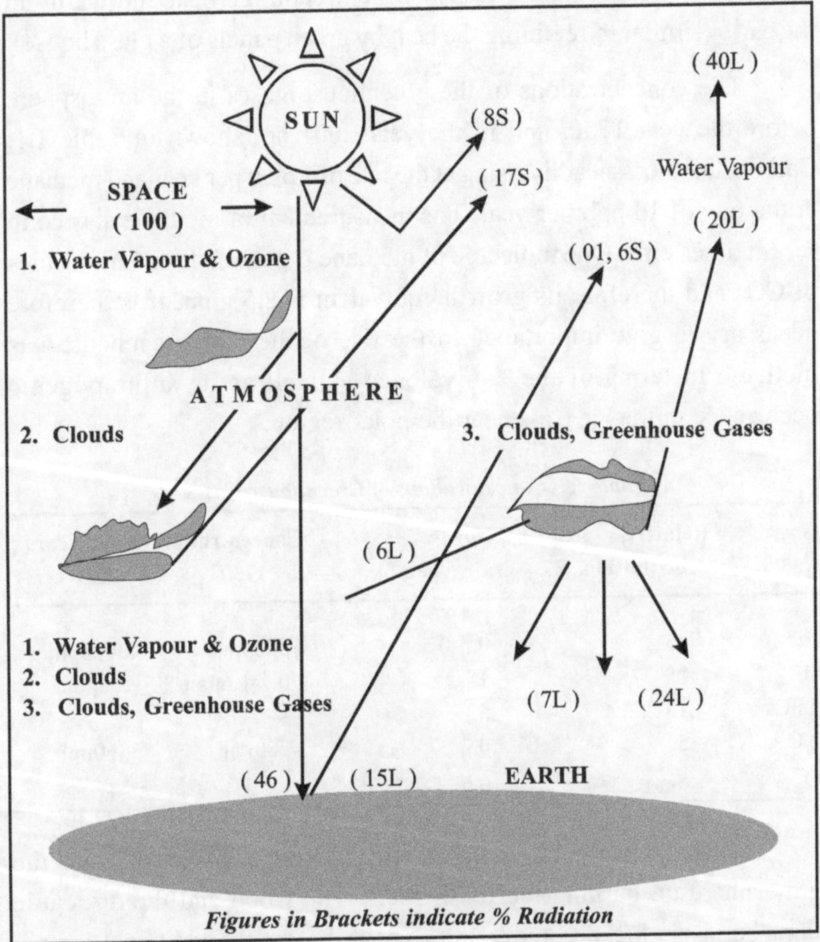

Fig. 1.4 Solar Radiation incident on Earth

Greenhouse Effect

Solar radiation incident on the earth is absorbed sufficiently over the surface. A portion of the energy is re-radiated into the space through a long wavelength radiation. This energy is then absorbed by the atmospheric gases namely carbon dioxide, nitrous oxide , methane, water vapour, CFCs (greenhouse gases) which retain the heat and warm up the earth (similar to retaining the heat by green panels of a glass house).

The concentrations of the greenhouse gases in the atmosphere before the year 1750, and in the year 2001 are shown in Table 1.1. Carbon dioxide is steadily rising at the rate of 1 ppm per year and methane at the rate of 10 ppb per year. The measurement of methane started in recent times only. One molecule of methane is generic for 25 molecules of CO_2 and therefore its growth equivalent is 2.5 ppm. It is therefore necessary to give importance to the rise of the new greenhouse gas methane in terms of the equivalence as well as its anthrapogenic occurrence in the ice caps near the polar region.

Table 1.1 Concentrations of Greenhouse Gases

Gas	Relative contribution, %	Growth rate /year, %	Concentration in air, years	
			1750	2001
CO_2	65	0.3-0.5	280ppm	360ppm
CH_4	25	1	0.7-1.6ppm	2.7ppm
CFCs	15-25	5	---	---
NO_x	5	0.2	280ppb	360ppb
O_3	8	0.5	---	---

Additional quantities of carbon dioxide are released into the atmosphere on burning the fossil fuels. The wood and the dry cattle dung burnt as a fuel in some countries contribute to the additional carbon. Also, the consumption of petrol, diesel, gas for motor cars and automobiles raise the level of greenhouse gases namely carbon dioxide and nitrous oxides. Coal and petroleum products refining dispense a lot of methane. Even methanogenic bacteria act on biomass and generate methane. Chlorofluorocarbons find their way into the atmosphere from refrigerants and fire fighting fluids. Oxides of sulphur enter the atmosphere through natural volcaonic eruptions as well as burning of fossil fuels. Marine sulphate reducing bacteria produce some quantity of hydrogen sulphide and emit into the atmosphere.

The global mean land surface temperatures have increased between 0.5-1.0 °F since the year 1900. During the last 15 years the world witnessed ten warmest years thereby getting the clue that global warming occurred.

Also the sea level has risen by 4-8 inches during the last century; about 25% of this rise out of the melting of mountain glaciers. The precipitation has increased by 1% on the globe with minimal rainfall around the so-called tropics. It is predicted that these figures may go upwards in the coming years.

Global Warming Potential

A 'global warming potential or value' is an estimate of the greenhouse gas emission in terms of Units of Millions of Metric Tons of Carbon Equivalents (MMTCE). In this respect HFCs (halofluorocarbons) and PFCs (perfluorocarbons) are the highest heat absorbing materials. Methane takes up heat 21 times as that of carbon dioxide molecule; nitrous oxide to the extent of 270 times that of CO_2. Even oxides of sulphur do absorb infrared heat but they are much short-lived unless present in abundance during volcono burst. It is therefore necessary to build an 'emission inventory' giving all these particulars.

Sinks for Greenhouse Gases

Some of the natural sinks of the greenhouse gases are: the *earth*, the *trees* and the *oceans*. A sink is a reservoir that accepts a chemical, or gas or compound from another type of cycle. All oxides of nitrogen, sulphur, chlorine, acids, hydrogen sulphide from atmosphere are taken up by the earth. Carbon dioxide is received by the oceans and their levels are even moderated. Plants metabolise carbon dioxide mostly and balances its presence in atmosphere.

Impacts of Global Warming

Global warming increases the surface temperature of the earth thereby changing to hot conditions. During hot days the atmospheric production of ozone is also more at ground. Ground ozone is deletereous to health of people. Owing to the hot conditions some of the diseases such as malaria, dengu fever and encephalitis manifest and take lives. The old persons experience cardiovascular problems and heat exhaustion.

On account of the increased heat in a zone, the evaporation of water bodies will be more; as a result of this the water levels in the lakes and rivers fall down. The navigation will be immensely affected due to shallow waters. Even recreation and tourism will be minimal under these circumstances. Moreover warm climate may extend the ice forming season also in the area. If the precipitation do not occur after the evaporation process in the same zone, the soil becomes very dry and do not render yields. Changes in the waterflow of rivers affect the hydro power plants. The power production statistics go array and cause hardship to the nation. Excessive evaporation of rivers and lakes also present the difficulty of fresh water supply to the people.

It is expected that the forests growth would be towards the cooler direction with the onset of global warming of atmosphere. A 3.6 $^\circ$F rise of atmospheric temperature over 100 years (that is expected now) is likely to shift forest towards the cooler northern direction of the globe at the rate of 2 miles per year. In the case of forests growing in high-latitude areas the extension of the trees in the existing tundra zone is predicted. The 'extended forest' growth no doubt serves as a sink for the carbon dioxide, but the excessive biomass that accumulate in the forest enhance the CO_2 levels in the atmosphere at the same time. Destruction of forest with the onset of fires is also anticipated in the event of global warming. Various bio-geographic models have been built and the studies indicated that the present desert land would increase by 185 %.

The wetlands on earth are to an extent of 4-6% of the land surface. It can be classified into (a) coastal wetlands and (b) inland wetlands. Coastal wetlands constitute predominantly marshes and swamps wherein tidal waters freely enter and go. The sea level rises very fast adjoining the marshes and erodes the coastal wetland over a time. Some portion turns out to be the dryland adjoining the marsh; however the area of the newly created dryland is not as large as the coastal wetland that was lost in erosion. Even flooding occurs during the stormy days in the dryland making it unsuitable for any marine activity. In the case of inland wetlands which are non-tidal, global warming may cause lower rainfall to occur thereby producing many drylands. These situations lead to draughts.

Further, it is expected that agriculture in the developing countries may be hampered on account of global warming. The prediction of climate, lack of technology and low adaptation to automation in agriculture become the causes for low productivity in under-developed and developing countries. With sufficient planning this catastrophic effect can be overcome.

The worst impact global warming is on the ecology of polar regions. Satellite data has shown that the snow cover in the arctic zone has declined by 10% since 1960. It has also indicated that the general temperatures in the polar regions have risen during the last millineum. The Alaska region warmed up by an average of $4\,^{\circ}F$ since 1950 and the thickness of the Pine Island Glacier of West Antartic Ice Sheet has come down by 5.2 feet. The Polar bears which live on the marine ringed seals were showing poor health on account of the non-availability of their prey. The sea ice that melts down in this area is causing hardship for the survival of the ringed seals; during spring months the melting of ice is rampant in this zone on account of global warming.

The fish also change their migration patterns depending on the warming of the region. Inland, coastal as well as ocean resident fish would be affected by global warming. The habitation of birds is largely determined by the existence of wetlands, beaches, tundra and marine species. Under the pleothra of global warming, many birds may prefer to land at the Northern hemisphere as the conditions are more congenial. It was discovered that the freezing of water vapour has been occurring only at greater altitudes as compared to the earlier situation in the tropical Tibet mountain range. It appears that the last 50 years were the hottest in the history of Tibet.

Sustainable Environment Through Control of Greenhouse Gases

If we limit the exit of greenhouse gases from the earth's surface into the atmosphere, it is possible to control the global warming. The outlook of industrial development throughout the world should be to achieve the productivity by pumping minimum quantities of carbon gases. The scope of this approach is mentioned in the following paragraphs.

Conservation of Energy

The fossil fuels, biomass and biofuels give energy and finally push carbon dioxide into the atmosphere. Eventhough efforts were put to replace these materials with clean fuels, no success has come. Therefore, the mankind has to conserve the energy by burning only limited organic fuels as may be required and using the energy very carefully. Energy audits at the source of generation and in the industrial sectors would also enable conserving the energy, thereby limiting the greenhouse gases into the atmosphere.

The transportation sector uses petroleum oils and gas to an extent of 33% in any country of the world. Enormous quantities of carbon dioxide is thus generated and this pushes the proportion of this gas to 85% of the total greenhouse gases. Their emergence is dependent on the 'travel growth', engine type, oil composition, occupancy by people and the weather conditions. If the travel planning is limited to necessary needs only within the municipal areas, one can cut down the carbon emissions. This is possible by allowing 'rush hour' schedules and restricting at other times. The composition of the fuel also determine the greenhouse gases such as sulphur dioxide, hydrocarbons, aldehydes etc. Depending on the type and efficiency of the auto engine, the quantity of emissions can be estimated and evaluated. It is pertinent in this connection that the authorities lay down regulations for running vehicles that are below 15 years old only and that too fitted with 4-stroke engines. The State should also encourage by insisting the occupancy of vehicles with at least 75% to its capacity than allowing them to run empty on the roads.

Alternative Approaches

The use of alternate fuels in the industrial and transport sectors has been fully recognised today throughout the world. The renewable solar, wind and ocean energies can be encouraged wherever possible; even some incentives may be provided to the energy suppliers and the users. Hydrogen is another fuel with least ramifications to cause ill effect in the environment. It is also abundant and easy to handle; therefore its usage in the form of fuel cells or directly or admixed with oil be encouraged.

Management of municipal waste yields rich dividends. First of all, the outlook shall be to prevent waste; next, a portion of it that accumulates can easily be recycled and supplied to the manufacturers for reuse. The profits of this scheme of using recycled materials be passed on to the ultimate users also. A good clean-up operation and reuse will definitely *minimise* carbon gases in the globe. It is imperative in this context that even wood burning in rural or forest areas should be discouraged.

Substitutes for the erstwhile CFC's should be used in all consumer products supplied by the industry. This will not only ensure controlling the greenhouse gases but also help protecting the earth from the exposure to excessive ultraviolet radiation that passes freely through the 'ozone hole'.

Methane Traps Under the Sea

So far the greenhouse gas, methane, is treated as one that is absorbing the infrared radiation emitted by the earth and contributing to global warming. Recent reports show that in many continents the gas is entrapped in marine sediments under the sea, as also super-saturating the sea waters. A gas hydrate of methane is a phenomenon in the Arctic zones. The gas hydrate is a crystalline solid looking like 'water ice' and consisting of methane molecules surrounded by a cage of water molecules. These materials are observed to be quite stable at ocean depths of 300 meters, and they cement loose sediments in the form of surface layer for many hundreds meters thickness. Estimates are that the carbon content of these hydrates are quite enormous and exceed that of the currently found fossil fuels on earth. Such a resource of methane gas hydrate was found by US Geological Survey Scientists off North and South Carolina shores. The ocean trough is a well-known offshore oil and gas area and at the basin the sediment is apparently 13 km. This large gas trap must be a good energy resource, leaving apart the potential for global warming. One important observation made by these scientists has been that these hydrate zones are prone for landslides in the Atlantic continental margin.

The ice canopy of the Alaskan continental shelf of Beaufort Sea between Harrison Bay to Camden Bay (about 250 km) was drilled and the water at depths ranging from 2.1 -38 m was collected and analysed for soluble methane gas. It was observed that many samples were super-saturated with methane. Similar ice caps with concentrates of methane were also noticed in the Alaskan and Canadian coastal waters of the Beaufort Sea. Part of the methane gas enters the atmosphere when the ice cap melts down and this feature is common during summer months.

Ozone Depletion

The border zone to the troposphere and the stratosphere has a thin lining of ozone layer. Ozone has been formed at this zone according to the following transformations:

1. O_2 molecules are photolysed into two oxygen atoms (slow);

2. O_3 (ozone) and O_2 are continuously being inter-converted as solar UV radiation passes through the ozone layer and breaks it.

Equations

$$O_2 \xrightarrow{h\nu} O + O \qquad \text{(oxygen at bottom atmosphere)}$$

$$O_3 \rightarrow O_2 + O \qquad \text{(ozone at stratosphere)}$$

$$O_2 + O + M \rightarrow O_2 \qquad \text{(ozone formation)}$$

$$Cl + O3 \rightarrow ClO + O_2 \qquad \text{(cleavage of ozone at stratosphere)}$$

$$ClO + O \rightarrow Cl + O_2$$

Bromine in oceans also contributes to depletion of ozone layer in a similar fashion to the extent of about 25%.

Pathways of Formation of Ozone at the Troposphere

Under the influence of high energy ultraviolet radiation falling on molecular oxygen, splits it into two atomic oxygen atoms. Each of these atomic oxygen react with one each of oxygen molecules and generate a total of two ozone molecules. The formation of ozone from oxygen and conversion back is a perennial phenomena, but with a reasonable balance of ozone in the stratosphere.

Down on the surface of the earth, we have a variety of organo-halogen compounds that are used in several syntheses. A category of Chlorofluoro carbons (CFC), usually in gaseous or liquid form have diverse uses ranging from sprays, refrigerants to fire fighting fluids. These substances are volatile and are not destroyed in the normal environmental cycle in the atmosphere. Their residence time is considerable in the atmosphere. With the drift of winds, the CFC's are carried to such heights as the stratosphere where they are decomposed into chlorine /bromine atoms by the incident ultraviolet light. It is this particular chlorine atom that reacts with ozone and destroy the thin layer. Several thousands of molecules of ozone are cleaved with a few atoms of chlorine. The net result is the depletion of ozone layer in the stratosphere.

It has been observed that each of the CFC's has a characteristic 'ozone depletiong power'. A majority of these are refrigerants, hair sprays, toiletries and fire retarding fluids. Since there is no mode of destruction of these chemicals on the earth, they are reaching the upper heights in this manner. Their high concentration in the Antarctica zone created a hole in the stratosphere and this is known as 'ozone hole'. The evidence for the ozone hole has come from the satellite pictures taken all round the globe.

The damage caused by the Ozone Depleting Substances (ODS) to the ozone layer is profoundly serious. The uv radiation from the sun could directly penetrate through the hole and cause health problems. Moreover, if this is allowed to extend, it may become catastrophic to the universe. In Table 1.2 are given the area figures of the hole. In Fig. 1.5 is shown the present scenario of 'ozone hole' as seen from a satellite.

Table 1.2 *Areas of Damage by Ozone Depletion*

Location	Area (thousands sq.km)
Australia	8923
USA	9362
Europe	10,498
Russia	17,078
Africa	30,335
N. America	25,349
Antarctica	13,340
S. Pole to 70 S	15,300
S. Pole to 65 S	23,890

Source : http://www.cpc.rcep.noaa.gov/products/stratosphere/sbuv2to/ozone_hole.html

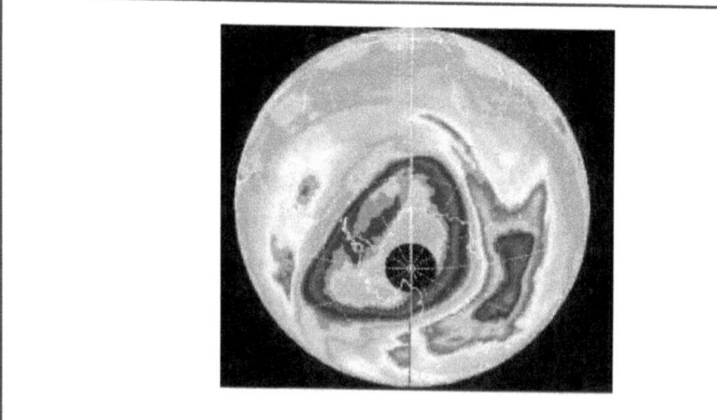

Fig. 1.5 *The Ozone Hole*

The first report of 'ozone hole' was made in the year 1954 by the British scientists. Since then there have been serious apprehensions on the continued use of CFC's. A Montreal Protocoal was evolved and according to this all developed countries should phase out the CFCs that are ozone depleting substances (ODS) by the year 2005 and the remaining countries by the year 2010. Since then research towards the development of non ozone depleting substances has begun and several products developed.

The characteristics of the CFCs and ODSs are shown in Table 1.3. Now, the biggest challenge to the scientific community and the administrators is to evolve methods for repairing the ozone hole in the atmosphere.

Table 1.3 *Ozone Depleting Chemicals and CFCs*

Chemical	Usage	Life time/ years	Approx. emissions in tons	Ozone depleting potential	Alternative candidates[#]
CFC-12	A/C, Refrigerators, Foams.	139	454	45	HFC-134a
CFC-11	-- do --	76	262	26	HCFC-123
CFC-113	Solvent	92	152	12	
CFC-114	Water Coolers, Chillers	---	---	---	HCFC-124
CCl4	Degreaser	67	73	8	
Methyl Shloroform	Solvent	8	522	5	
Halon-1211	Refrigerators	22	79	0	
Halon-1301	Fire extinguisher	101	3	4	
R-502	Refrige-mix	---	---	0.30	HCFC-22

Sources: Botkin and Keller, "Environmental Science", 1995.

 # Albert Thumann and D.Paul Mehta, Handbook of Energy Engineering, 1994.

Acid Rain

The acid rain is a menace to any country situated on the globe. High industrialisation within or in the neighbourhood of the country brings in this effect. The sulphur and nitrogen oxides emanating from the burning of fuels either in power plants or refineries or on road transport mix up with water in the atmosphere, and the rain that follows in the zone brings down these species onto the land in the form of acid rain. Acidic dew and snow are other manifestations of acid rain. The rain is acidic in character and it has been concluded that the pH of rain water if lower than 5.5 on the scale is positively acid rain. Acid rain formation is mostly man-made, but in some natural disasters such as earthquakes, volcano eruptions and hot water spring's sprays, a few oxides of sulphur emerge out and form acid-aerosols in the atmosphere.

The sulphur oxides are essentially formed during the burning of fossil fuels in the thermal power plants. Here the coal or oil is usually burnt to give steam. These fuels contain sulphur anywhere between 0.5-4% depending on the location from where the fuel is mined or drilled. About 75% of the ingredients of acid rain accumulate from these activities. In addition, domestic use of fossil fuels, refinery processes, road transport uses and shipping add on the remainder portion of the acidic components in the atmosphere. A second menacing gas that produces acid rain in the atmosphere is NOx. A major chunk of this comes from the high temperature burning processes, to the extent of about 50% especially from the engines of automobiles on road. The burning of fossil fuels in the thermal power stations also contribute about 27% of these oxides. Soil putrification by natural bacteria also push a few nitrogenous gases into the atmosphere which further degrade into NOx.

Acid rain is a perennial feature in Europe, which has many small countries closely situated. The acidic components generated in one country reach the other countries in the form of acidic precipitation after being driven away by winds. Similarly the winds from the Canada side push the acidic pollutants of the atmosphere into USA and cause acid rain in N. America.

Some of the remedial measures which focus mainly on limiting the formation of sulphur oxides that make acid rain are:

1. Water-washing of coal to remove sulphur compounds prior to its use in furnaces.

2. Flue Gas Desulphurisation (FGD) by passing the flue gases coming out from a coal-fired power station through filters containing calcium carbonate slurry.

3. Use of Pressurised Fluidised Bed Combustion (PFBC) in which air is forced through a bed of coal and limestone.

Acid rain causes corrosion of metals and metallic structures. It kills the plants and seriously affect the soil too. The forest wealth of a country is destroyed at one stroke. The stone monuments, sensitive and heritage buildings also get damaged. Finally the run–off acidic components, as well as acidic precipitation causes ecological damage to the lakes, estuaries and rivers. Even the natural energy resources such as forests, rivers, lakes, the food chain on earth and the minerals are significantly affected by the incidence of acid rain.

Questions

1. What are the typical parameters that determine a sustainable environment? Discuss how the eco-system is responsible for creating various types of energy resources.

2. Explain the photosynthesis in plants.

3. Discuss the energy balance of earth.

4. Explain the Greenhouse effect. Which are the greenhouse gases and how do they transform the atmosphere around the earth?

5. Explain the global warming feature and focus the impacts of this phenomenon on the disturbance to the sustainability of environment.

6. Give an account of the existence of methane traps under the sea and contemplate the ways of bringing this energy resource over the earth's surface for meeting the energy demands.

7. Write down the scheme of formation of ozone at the interface of troposphere and the stratosphere and explain how this ozone is depleted at the present juncture. Give an account of the remedial measures to be adapted by nations of the world.

8. What is acid rain? How does acid rain damage the environment? Narrate the steps to be taken to mitigate the damage.

9. Explain how volcanoes balance the energy systems on the earth and its atmosphere? Do you consider that volcanic explosions are advantageous or disastrous to the mankind?

2

ENERGY, WORK AND POWER

In our daily life we use energy to get the work done. Work is thus the resultant of application of energy, may be that of heat, electric power or chemical. Human body exerts energy from the breaking of the carbon-hydrogen-oxygen chemical bonds of the muscles. The muscles attain the C-H-O type of bonds from the foods eaten and nourished by the human body. The gases released in certain chemical reactions spontaneously move the machine parts thereby causing work. Burning of petroleum or natural gas in a set-up of cylinder and piston renders work. A speedy wind moves the turbine blades in a 'wind mill'. The water falling down from a height moves the turbine blades and generate electricity. Thus, the performance of work in an energy system is an important feature.

Production of energy and its utilisation by men in different societies of the world varies according to the geographic location. In cold regions of the world people require more energy. Also the developed nations consume more energy as compared to the developing countries.

A comparison of these has been made in Table 2.1. It is seen from the table that the utilisation of energy in developed countries is ten-fold larger than that in developing countries.

Table 2.1 *World Primary Energy Consumption (1998)**

Country	Quantity, Quadrillion Btu
USA	94.57
China	33.93
Russia	25.99
Japan	21.28
Germany	13.83
India	12.51
Canada	11.85
France	10.00
United Kingdom	9.75
Brazil	8.08
Total incl. others	377.50

** Primary Energy includes Petroleum, Natural Gas, Coal, Hydroelectric, Nuclear, Geothermal, Solar, Tidal and Biomass*
Source: *Energy Information Administration, US Department of Energy Database,*

June 2000

Although several classifications of the types of energy in terms of renewable, non-renewable, exhaustible, non-exhaustible, conventional, non-conventional etc. have been made by many authors, we prefer the use of conventional, non-conventional words for energy in this book. Under this category, the energy obtained from fossil fuels namely, petroleum, coal, gas and oil shale, hydro and nuclear sources constitute conventional energy. Solar, wind, tidal, wave, and bio energies fall under non conventional type. As the energy from fossil fuels is expected to be depleted from the world soon, the non conventional energies were rapidly tapped and augmented to the conventional type. Utilities such as town heating, lighting, transport and remote mountain and space applications adapt both types of energy. Excess energy if produced from any of the non conventional type can be stored by suitable techniques which have been described in the subsequent chapters. The present scenario of production of primary energy in the world has been shown in Table 2.2.

Table 2.2 *World Primary Energy Consumption (1998)*[*]

Country	Quantity, Quadrillion Btu
USA	72.55
Russia	41.34
China	33.13
Saudi Arabia	21.00
Canada	17.18
United Kingdom	11.67
India	9.95
Iran	9.89
Norway	9.39
Mexico	9.29
Total incl. others	382.00

Source: *Energy Information Administration, US Department of Energy Database, June 2000*

By now we visualise energy as existing in several forms and undergoing transformations in order to perform different types of work. The primary energy forms are: fossil fuels, solar, wind, hydro and geothermal types whereas the secondary energy forms are electricity, sound, steam etc. The primary forms of energy which are very familiar to us cannot be applied for performing the work directly in some cases and it has to be converted into a secondary form for effective application. Thus the electrical power came into prominence.

In a system which has a source of energy and the surroundings that accept the effects caused by the source, physical observations on the surroundings reveal the extent of application of energy. Consider a test tube having sodium hydroxide solution placed in a beaker containing ice. When an acid solution is added to the contents of the test tube, heat develops rapidly. On absorbing this heat, the surrounding ice in the beaker starts melting. Here the source is the chemicals in the test tube and the surrounding is the ice in the beaker. This is a case of non adiabatic system, that is, a set-up in which heat generated is wholly equilibrated between the source and the surroundings. We have another set-up in which the heat energy of the source does not get exchanged with the surroundings.

This is adiabatic system which is thermally insulated from the surroundings so that heat does not get exchanged. The hot water which is stored in a thermos flask is an example of an adiabatic system. Further, chemical reactions performed in the laboratory have been categerised into three types, namely open, closed and isolated, all of which are pictorially shown in Fig. 2.1.

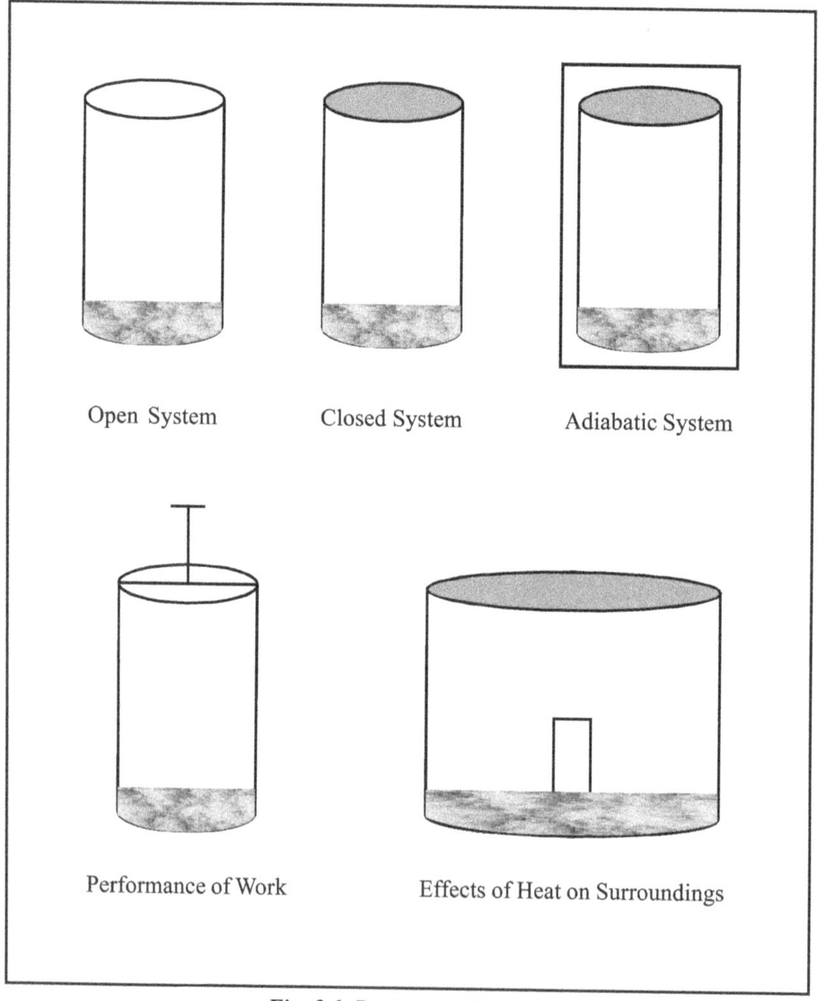

Fig. 2.1 *Depiction of Energy*

When energy is transferred to the surroundings some physical movement occurs and it results in the performance of what is called work. The displacement of piston of an engine performs horizontal motion of the whole body thereby performing work. A horse pulls a cart; an elephant lifts wooden logs; a fly-wheel moves the vehicles on the rails; the piston of a cylinder in which calcium carbonate is decomposed is lifted up. Based on all these examples, we can deduce that work is the product of distance and force.

$$\text{Work} = \text{distance (height)} \times \text{force.}$$

$$W = d \times ma$$

$$\text{or} \qquad = h \times mg$$

where   a is the acceleration, and

       g is the acceleration due to gravity of earth .

Thus when an object is moved by one foot applying a force of one pound, the work done is equal to one foot-pound. If an object is moved by 10 feet with a force of 20 pounds, the work done is equal to $10 \times 20 = 200$ foot-pounds. If we introduce the time factor in the process, say seconds/minutes/hours when the energy transfer takes place, the work done can be expressed as foot-pounds per second/min/hr as the case may be. This by definition is termed as power, which is the *rate* at which work is done. In the field of 'dynamics' the usage of power is very frequent. The term 'horse power' (HP) is commonly used and it represents that 550 foot-pounds of work was done in one second in a system. It is a measure of any power machine. When a crane lifts 550 pounds of object to a height of 10 feet in one second, it delivers a work of 5500 foot-pounds per second and the power of the engine is said to be 10 horse power.

Common Units

1. *Calorie :* Amount of heat necessary to raise the temperature of one gram of water kept at 15 degrees Centigrade by one degree centigrade.

    $$1 \text{ calorie} = 4.184 \text{ Joules;}$$
    $$1 \text{ therm} = 105.5 \text{ MJ}$$
    $$1 \text{ joule} = 10^7 \text{ ergs (old unit)}$$

In nutrition science kilo calories are used.

2.   ***BTU :*** British thermal unit is the amount of heat necessary to raise the temperature of one pound of water at density 1 by one degree Fahernheat.

$$1 \text{ BTU} = 251 \text{ calories}$$
$$= 1050 \text{ joules}$$

Both calorie and Btu are heat energy units.

3.   ***Watt*** : measuring unit of power, especially electric power; represented by symbol 'W' for electricity.

$$1 \text{ watt} = 1 \text{ joule/sec} = 10^7 \text{ ergs/sec.}$$

4.   ***Horse Power (HP) :*** unit for engines performing work;

$$1\text{HP} = 550 \text{ foot-pounds/sec};$$

also $1 \text{ HP} = 0.75 \text{ kw.}$

5.   ***Kilo - Watt - Hour (KWH)*** : The power supply of 1000 watts for one hour; also equivalent to 36,00,000 joules ie. 3.6 MJ.

$$K = \text{kilo} = \times 10^3$$
$$M = \text{mega} = \times 10^6$$
$$G = \text{giga} = \times 10^9$$
$$T = \text{tetra} = \times 10^{12}$$
$$P = \text{penta} = \times 10^{15}$$
$$E = \text{exa} = \times 10^{18}$$
$$1 \text{ exa joule} = 10^{18} \text{ joules} = 1 \text{ Quadrillion (Quad)}$$

Other Units

1.   toe = ton oil equivalent = 44.8 g j = 10,500 m cal.

2.   toc = ton coal equivalent ; 1 toe= 1.5 toc.

3.   Ev = electron volts;

E commonly used in nuclear energy;

$$E = 1ev$$
$$= 1.602 \times 10^{19} \text{ joules}$$
$$= 1.602 \times 10^{-12} \text{ ergs}$$
$$= 6.7 \times 10^{-19} \text{ calories.}$$
$$Mev = 1 \text{ million electron volts.}$$

Salient Information

1. Man's capable energy = 1/10 – 1/18 HP;

2. Automobile's rendering energy = 1 – 50 HP(2,3,4 wheelers)

3. Train's mobile power = 6000 HP.

4. A large modern wind turbine is rated at 500 kW.

5. A large modern fossil fuel-run power plant is rated at 1 GW (1000 MW).

Questions

1. Define the terms: Energy, Work and Power, and discuss with one example each.

2. Explain how energy is transformed into work.

3. Define: calorie, BTU, horsepower and watt.

Fill up the Blanks :

1. The radiation from Sun onto Earth gives _______________ and _______________.

   (electricity and magnetism)

   (light and heat)

   (precipitation and wind)

2. The ultimate result of any applied energy on a system is _______________ and _______________.

   (weight and loss)

   (heat and work)

   (noise and radiation)

3

ENERGY THERMODYNAMICS

In this universe we come across different types of energy systems and we harness each one of them for our benefit. At one stage all these types of energies get linkage to heat energy and therefore the fundamental laws of thermodynamics govern these processes. The two laws of thermodynamics pertinent to energy systems are the first and second laws.

FIRST LAW OF THERMODYNAMICS

This is also known as the *Law of Conservation of Energy* states that the sum total of all the energies in the universe is constant. In course of time a particular type of energy might undergo transformation to a different type, for example, light converted to heat; chemical energy transformed into sound and light etc. But the sum total of all these energies is maintained constantly. In other words energy is indestructible in the universal operations.

If we consider a closed system having well-defined boundaries and shape and that not allowing any transfer of energy through the system, and we perform a reaction within it, then the algebric sum of different energies generated inside in toto contribute to the total energy within the system.

Whenever heat energy (dQ) is added to a system, it is distributed in the system as an increase in internal energy (dU) and part of work (dW).

Mathematically,

$$dQ = dU + dW$$

The effects of the energy provided to the system are pronounced more on systems of small magnitude in the form of increased temperature. For the quantum of heat Q a significant amount of heat energy (U) absorbed by the atoms that underwent some transformations and simultaneously work W was performed; hence the percent efficiency of the work done can be mathematically written as:

$$E = \left(\frac{W}{Q} \right) \times 100$$

The first law of thermodynamics does not specify the direction of the energy although energy processes take place in specific directions, for example, heat flowing from hot body to a cold body. Also, the law does not mention the heat losses in the operations, which explain the low efficiency of the system. Some of these anamolies are answered by applying the second law of thermodynamics.

SECOND LAW OF THERMODYNAMICS

The law states that the various forms of energy differ in quality; some are more useful in rendering work whereas others are less-useful, for example electrical energy renders more useful work than say solar or wind (mechanical) energy. The input and output forms of energy determine the efficiency of energy conversion, for example, electricity-electricity conversion is very high as compared to electricity-heat or electricity-light or heat -work as well as heat-electricity conversions.

Some heat loss occurs in every energy conversion mentioned above and the heat goes as internal energy as well as radiation into the environment and friction, but not as work. Therefore all systems which work on the intake of energy would never work with cent per cent efficiency. No heat engine, however well conceived and built, will not work to cent percent efficiency; but the efficiency is based on the difference of input temperature (heat) to that of the output heat operating in the system. Thus all machines employing heat, eg. Heat engines or refrigerators, operate following the second law of thermodynamics. This sequence has been explained below in Fig. 3.1.

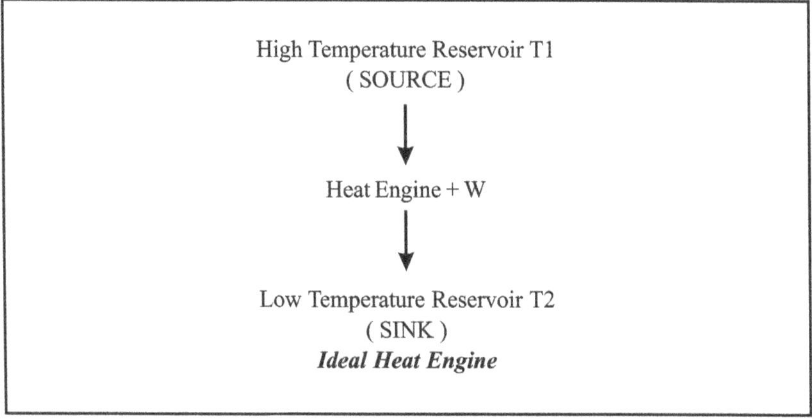

Fig. 3.1 Concept of Heat Engine & Refrigerator

If we consider that W is the work delivered on absorbing heat at high temperature source which is maintained at a temperature (T1), the heat equivalent is Q1; the efficiency of this operation is

$$E = \left(\frac{W}{Q1}\right) = W/Q1$$

According to second law of thermodynamics, always some heat is rejected in the operation. The work delivered is thus

$$W = Q1 - Q2$$

of heat energy.

Substituting this work equation in the efficiency equation above,

$$E = \frac{(Q1 - Q2)}{Q1}$$

and instead of heat units of high end and low end reservoirs the absolute temperatures are considered, the efficiency equation becomes,

$$E = \frac{(T1 - T2)}{T1}$$

i.e., $$= \left(1 - \frac{T2}{T1}\right)$$

In terms of per cent efficiency of the engine, the equation can be represented as

$$E\% = \left(1 - \frac{T2}{T1}\right) \times 100$$

It follows from the second law of thermodynamics that one has to maintain a large gap of temperature between high temperature reservoir to low temperature reservoir in order to achieve high efficiency. In a steam engine, if steam at an absolute temperature of 550 K is entering and performing the work and if the resultant temperature of the outlet steam has fallen down to say 300 K after the work was done, the quantum of heat utilised was 550-300 = 250 K; then the efficiency of the steam engine shall be

$$E\% = \left(1 - \frac{300}{550}\right) \times 100$$

$$= 45.45$$

However if the input temperature of the superheated steam is raised to 750 K and the outgoing steam temperature remain at 300 K the percent efficiency of the engine shall be

$$E\% = \left(1 - \frac{300}{750}\right)$$

$$= 60.00$$

Free Energy

Of the total heat which is converted into work in a system, a small component of it was used for the expansion of the material, and the remainder performed the so-called useful work through electrical or mechanical conversion. Actually this available work is important and is termed as 'free energy'. This is measured at constant pressure and is denoted by the symbol G, which indicates a change in state of the system. From theoritical considerations, if G becomes a −ve sign, the process in the system is spontaneous; otherwise, it happens non-spontaneous. The spontaneous energy conversion processes are invariably irreversible. All irreversible reactions do not return to original state. After completion of an irreversible process it is impossible to restore original condition of the system and surroundings. We find in the processes that a few types of energies that are not of high quality after rendering useful work allow dissipation of the energy to the surroundings in the form of friction, heat flow, throttling etc. The processes involving conversion of heat to work are not complete conversions, as a portion of heat is always lost. Work can be completely converted to heat. Electricity also can be fully converted to work. In a spontaneous process, permanent degradation of energy occurs resulting in a lot of disorder with a loss of available energy. All these concepts of thermodynamics are to be understood to follow the energy conversions and evaluate the work performed as a result of utilising the available energy in the system.

Questions

1. Explain the first and second laws of thermodynamics as related to energy systems.Explain Carnot equation and efficiency.

2. Write notes on 'free energy'.

4

FOSSIL FUELS

Fossil fuels are the most important conventional energy sources in the world. They constitute coal, petroleum and natural gas. Like any natural resource, constant utilisation of these materials by the mankind depletes these very fast. It is feared that the resources may be just sufficient for a century from now henceforth. These resources are having the energy stored in the carbon- hydrogen chemical bonds; and when they are broken by burning, the energy is liberated. All these are organic materials found in the earth. These are the stock materials for the production of synthetic fuels and chemicals. The global distribution of coal is to the extent of 22%, petroleum to 40% and natural gas about 24%; and all the three put together comes to around 86%. The fossil fuels render primary energy, namely heat, on burning in air. The civilization knows the use of coal since the year 1850 AD, crude oil since 1875 and natural gas since 1900. The total production and consumption of coal, petroleum and natural gas in India is shown in Table 4.1.

Table 4.1 Production and Consumption of fossil fuels in India (1999)

Type	Production	Consumption
Coal*	326.06	336.68
1. lignite	25.41	---
2. bituminous	300.66	---
3. anthracite	na	---
Petroleum**	747	1,930
Natural gas***	0.750	0.752

*   *in million short tons during 1977*

** *in thousands bbl/day; includes crude oil, natural gas, other liquids & refinery processings.*

*** *dry natural gas in trillion cu.ft. (tcf) during 1999.*

The Total Primary Energy Production (TPEP) and Total Primary Energy Consumption (TPEC) for India during the year 1998 is shown in Table 4.2. Eventhough the fossil fuels are filling the gap of energy shortage to a larger extent for India, alternate energies are to be augmented to obtain an exact balance of energy requirement.

Table 4.2 Total Primary Energy Production (TPEP) & Total Primary Energy Consumption (TPEC) in India, 1999

TPEP		TPEC		Deficit	
Quads	**Mtoe**	**Quads**	**Mtoe**	**Quads**	**Mtoe**
9.0	227,5	11.6	293.0	2.6	65.5

1 Quad = 25.2 Mtoe

Source: *DOE from: http://www.fe.doe.gov/international/indiover.html*

COAL

Coal was the best fuel during the nineteenth century in Europe. It was catering the energy needs to the extent of 80% in that continent. Extensive mining and subsequent consumption were made at that time, and in the 20th century, its consumption has fallen down to nearly 45% and these days it is to the level of 30% only.

The world reserves of coal are about 6000 billion tons of which only about 3% is so far mined. India's proven coal reserves as of January 2000 stand at 82,396 million short tons. The country largely depends on this fossil fuel for energy generation. The production of coal in India during the year 1999 is 328 million short tons whereas the consumption was 348 million short tons in the same year. The coal of Indian origin has high ash content (35%) and hence requires washing. However coal continues to be the dominant fuel for energy generation and production of electricity on account of its abundance. Coal is abundant in USA, Russia, central European countries, China and India.

Coal is interstratified in the earth layers along with limestone, clay, sandstone and shales. Originally the plants which were grown on the earth go into low-lying areas and swamps after their death in the hot humid regions. They were subjected to the pressure from the top layers of muds and also to microbes. Over a time of about two to three hundred million years under these conditions, the organic bodies get converted into a brown-black mass in the first stage of transformation, called 'peat'. Peat bog or fen is a low carbon coal, which subsequently gets converted into another type known as 'lignite'. After many metamorphic changes, the lignite converts to a lumpy ignitable mass called 'coal'. Coal is a very highly complex organic mixture of ignitable compounds. It can be pulverised, broken into small pieces or shaped. It has high calorific value if burnt and therefore serves as an excellent fuel.

PEAT

When plants submerged under the earth in swamps over millions of years get metamorphically transformed by heat, pressure and microbiological activity into complex organic molecules, the first stage of material that is formed is the peat. The former USSR, Finland and Iceland have large peat fields. It is brown in appearance and possess a lot of water(80-90%). Upon drying, a solid porous mass is formed. The composition of the dried mass is given as: C = 57%; H = 6.1%; O = 34.9% and minerals, 2%. Peat in most dried varieties contain ash in the range 2-10%.

The usual calorific value of peat is around 5000 Btu/lb or 18 kj/kg (compare wood, 20kj/kg). It is shaped into briquettes and used directly in locomotives. In cold countries, dried peat is used for town heating. It is also subjected to destructive distillation to afford hydrocarbon oils, fuel gases, wax and tar.

LIGNITE

This variety of coal is formed by further metamorphic changes of peat in the earth. In India, lignite belts are located in Tamilnadu (Neyveli), Rajasthan (Palan) and Assam. Lignite is an amorphous dark brown mass without any lustre. Originally this material too has high moisture content(30-50% as it comes out from earth). It has about 30% of volatile organic matter and burns with sooty flame. It has low calorific value, about 6000-7500 Btu/lb or 14-19 kj/kg. It finds extensive use in the boilers and heating pans. A huge chunk of lignite is utilised in power generation at the locations where it is mined, as the transportation of the material is uneconomical due to the water and volatile matter present in it.

Coal

The final form of the buried plant and vegetation under the earth's heat and pressure over millions of years results in stratified coal fields. Many reserves of coal are found in Indian states of Bihar, Orissa and Andhra Pradesh. Coal is a highly complex mixture of organic compounds with several nitrogenous and sulphur compounds. There are four forms of coal, each distinguished by the fixed carbon content. These are: sub-bituminous coal, bituminous coal, semi-anthracite and anthracite. All bituminous coals look like bitumen and burn with smoky yellow flame. They appear as black layers and possess bright texture. They break into blocks and can be shaped also. The calorific value is in the range 10,330-13,150 Btu/lb or 21- 36 kj/kg. A typical sub-bituminous coal has the proximate composition as follows: moisture, 13.9%; volatile matter, 34.2%; carbon, 41.0%; ash, 10.9% and a bituminous coal has moisture, 5.9%; volatile matter, 43.8%; carbon, 46.5%; and ash, 3.8%. These materials contain sulphur and nitrogen as well, which can be removed by washing.

The anthracite type is very hard, compact, black-lustered coal possessing high fixed carbon content. It gives 13,000-15,000 Btu/lb or 30-36 kj/kg of heat which is the highest among the coals. This variety also contains sulphur and nitrogen compounds in addition to ash. A typical anthracite has the proximate composition: moisture, 4.4%; volatile matter, 4.8%; carbon, 81.8%; ash, 9.0% and calorific value, 13,130 Btu/lb. Most of the Indian varieties of coal have low fixed carbon and hydrogen (almost half of USA variety) and high ash and sulphur (3 times more than the USA variety) and on account of this variation, the fuel value of Indian variety is low. This is prominently displayed in the overall performance of an Indian coal-based thermal power plant (that runs at 20-30% efficiency) as compared to that of a similar US plant that runs at 35-40% efficiency. The physico-chemical data on all types of coals is shown in Table 4.3.

Table 4.3 Physical Properties of Varieties of Coal

Property	Peat	Lignite	Bituminous Coal	Anthracite
Appearance	Brown-black	Brown	Brown-black	Black
Hardness	Porous	Soft	Soft	Hard
Water content, %	65	37	6	5
Ash %	---	5	4	5
Sulphur %	---	0.6	3	0.8
Carbon %*	57	41	47	82

** On Dried Basis*

Destructive Distillation of Coals

The process of subjecting coal to thermal treatment at high temperatures with or without air results in the formation of a very hard mass called 'coke' together with a few fuel gases. This process is called destructive distillation of coal, or carbonisation or coking. During the carbonisation of bituminous coal at low temperatures between 850-1300 °F semi-coke, many liquid products and a few gases are formed. The coking rate is rather slow and the liquids are motor fuel, water and tar. The gases are methane, ethane, carbon oxides, ammonia and nitrogen.

During high temperature carbonisation above 1650 F, more of the same gaseous products are formed. The hard coke results in abundant quantity which is used in metallurgy. A small portion of the liquid oil undergoes aromatisation yielding significant quantity of coal tar. On the whole, the main products of destructive distillation of coal affords the following categories of chemicals:

1. Coal gas, a mixture of gaseous hydrocarbons (useful as fuel)
2. Coke, a hard solid mass (useful in metallurgy and in graphite manufacture)
3. Coal tar, a raw material for coal chemicals.
4. Ammoniacal liquor, a raw material for ammonia.

The general reactions that take place during the gasification of coal or coal products can be written as:

$$Coal \rightarrow CO, CO_2, H_2, CH_4 + C \text{ (carbon)} \qquad \ldots\ldots(4.1)$$
$$C + H_2O \rightarrow CO + H_2 \text{ (endothermic)} \qquad \ldots\ldots(4.2)$$
$$2C + 3/2O_2 \rightarrow CO_2 + CO \text{ (exothermic)} \qquad \ldots\ldots(4.3)$$
$$CO + H_2O \rightarrow H_2 + CO_2 \text{ (exothermic)} \qquad \ldots\ldots(4.4)$$
$$CO + 3H_2 \rightarrow CH_4 + H_2O \text{ (exothermic)} \qquad \ldots\ldots(4.5)$$

Coal Tar Distillation

Coal tar is a viscous oily liquid, black in appearance and smells aromatic. It is an important product obtained during the destructive distillation of coal. In fact, coal tar is the source for about 300 industrially useful chemicals, amongst which benzene, xylene, naphthalene, phenanthrene, anthracene, phenols, cresols, pyridine are very prominent. Fractional distillation of coal tar is performed in cast iron stills connected to fractionation columns and receivers. As the dehydrated coal tar is subjected to heating, characteristic fractions of the oils distil out and are collected in receivers. In Table 4.4 are shown the various fractions, their boiling range, and the constituents of these fractions. Individual compounds from each of these fractions can be obtained by subjecting them to extensive fractionation.

Table 4.4 *Coal Tar Distillation Products*

Fraction	Boiling Range°C	Products	Uses
Light oil	200	Benzene, Toluene, Xylene	Solvents, Synthesis
Middle Oil1	200-250	Tar Acids, Naphthalenes, Pyridine	Synthesis
Middle Oil2	Above 250	Phenanthrenes, Anthracenes, Quinolene	Synthesis
Residue	---	Pitch	Road Laying, Tiles

Water Gas

When steam is passed over a red hot bed of coke maintained at a temperature of 1000 °C, a gas is generated. The components of the gas are hydrogen and carbon monoxide. As the products are formed the temperature slowly falls down indicating that the reaction is endothermic. Kinetic measurements have established that the reaction occur according to:

$$H_2O + C \rightarrow CO + H_2 - 29 \text{ kcals.}$$

If the lower temperature is allowed to continue, the course of reaction takes a different turn resulting in the formation of abundant quantity of carbon dioxide instead of carbon monoxide. Since carbon dioxide has no calorific value, it is an undesirable product. Hence, the reaction between steam and coke has to be manipulated at a rather higher temperature, namely 1000-1050 °C in order to obtain the two calorific substances namely carbon monoxide and hydrogen. The decomposition of water under these conditions is to the extent of 93% and the composition of the product is: H_2 = 50.7%; CO = 48%; CO_2 =1.3%. This gas is known as 'water gas', and its calorific value is 280-310 Btu/cu.ft. The calorific value of water gas can be enhanced by blending gaseous hydrocarbons and the resultant blend is called 'carburetted' water gas.

Producer Gas

Inferior types of coals such as peat, carbonaceous tars etc. are placed in a specially designed furnace called 'producer' and heated to 1000-1400 $^{\circ}$C. Immediately hot air is forced into this mass, when a sequence of pyro reactions occur resulting in the generation of a mixture of CO, CO_2, H_2 and N_2.

$$C + O_2 \; = \; CO_2 + 97000 \text{ Cals (in lower parts)}$$

$$CO_2 + C \; = \; 2CO - 39000 \text{ Cals (in upper parts)}$$

The composition of the gas is:

$$CO \; = \; 22\text{-}30\%$$
$$H_2 \; = \; 8\text{-}12\%$$
$$N_2 \; = \; 52\text{-}55\%$$
$$CO_2 \; = \; 3\%.$$

This is called ' producer gas' and it has a calorific value of 1300 k Cals/m^3.

Analysis of Coal and Coal Products

The methods of Proximate and Ultimate Analysis of coal and coal products are specified in ASTM and BS Specifications, briefs of which have also been dealt in Encyclopedia of Chemical Technology, Vol.16, 3rd edition, pp.270-272.

PETROLEUM

Petroleum is a key material in the generation of light, heat and mechanical energy. The reserves of oil are distributed throughout the world, both on land and offshore. In modern era, mining of petroleum gained a lot of importance as it is a direct source of energy. It is easy to handle this material in industry. The countries, Saudi Arabia, Kuwait, Iran, Iraq, Venezuela, Russia, Mexico, USA and India produce this oil through processes of exploration from nature. India's domestic production of petroleum in the wells of Bombay High, Upper Assam, Krishna-Godavari-Kaveri basins and Cambay put together was 747,000 bbl/ day during the year 1999.

The country also imports huge stocks of oil and its products from foreign countries every year. The crude oil thus obtained is transported from sea ports over land or seas to 22 oil refineries, distilled into several fractions and processed into various useful products. Nearly 30 % of energy share is contributed by petroleum in the country. The world's energy demand from the oil resources is billed at 4 billion tons oil equivalent (b toe) per annum.

The origin of petroleum is similar to that of coal. It was formed in the earth's bed on land or seas consequent upon the decay of dead animals and plants. These bodies settled down to the bottom of the seas and were covered with sediments and rocks. They remained as swamps for some millions of years. Due to limited oxygen supply in the sediments, and least circulation of water, most of the organic matter remained intact. More organic matter formed thus continued to be piled up over the earlier layers of the sediment. The microorganisms in the earth attacked these materials and partially decomposed them into petroleum. The excessive heat at the site resulted into decomposition of this organic material into the 'natural gas'. The petroleum thus formed got embedded into porous rocks around. The redistribution of the oil took place within the rocks by agglomeration of droplets and 'pools' of petroleum were thus formed (Fig. 4.1).

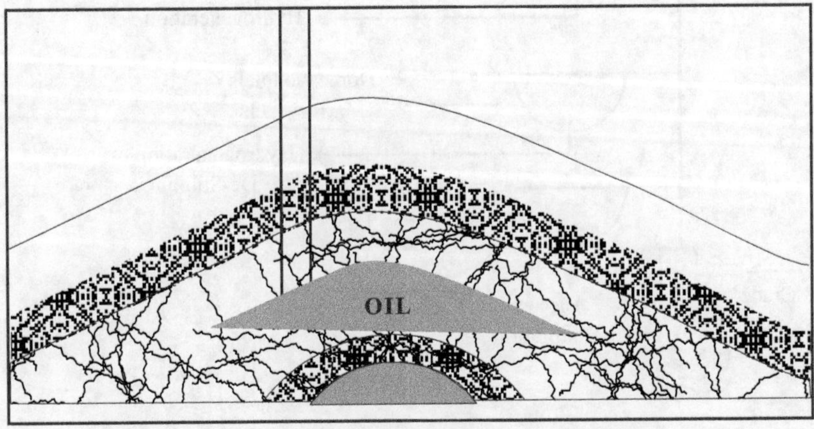

Fig. 4.1 Formation of Petroleum in the Earth & Drilling

The geophysical exploration of oil, drilling, separation of gas, transportation of crudes and refining processes are   high technology aspects. Even the blending of additives, and development of synthetic materials from petroleum and environmental pollution studies on land and seas (marine oil spills) are grey areas in science and technology.

Refining Process

The crude oil is placed in a distillation vessel fitted with fractionation columns, condensers and receivers. On heating the liquid gently, the low boiling gaseous fraction distils out and is thus separated. This is followed by light oils, middle distillates, wide-cut gas oils and finally residual oils. The low boiling gases separated are methane, propane and butane in the boiling range of -15 to 50 $^{\circ}$C. Light petroleum comprising of gasoline, solvents such as hexane, heptane, octane, nonane etc. are separated within 50 to 100 $^{\circ}$C range. The middle distillates yield kerosene, diesel, jet fuel etc. in the temperature range of 100 to 300 $^{\circ}$C.

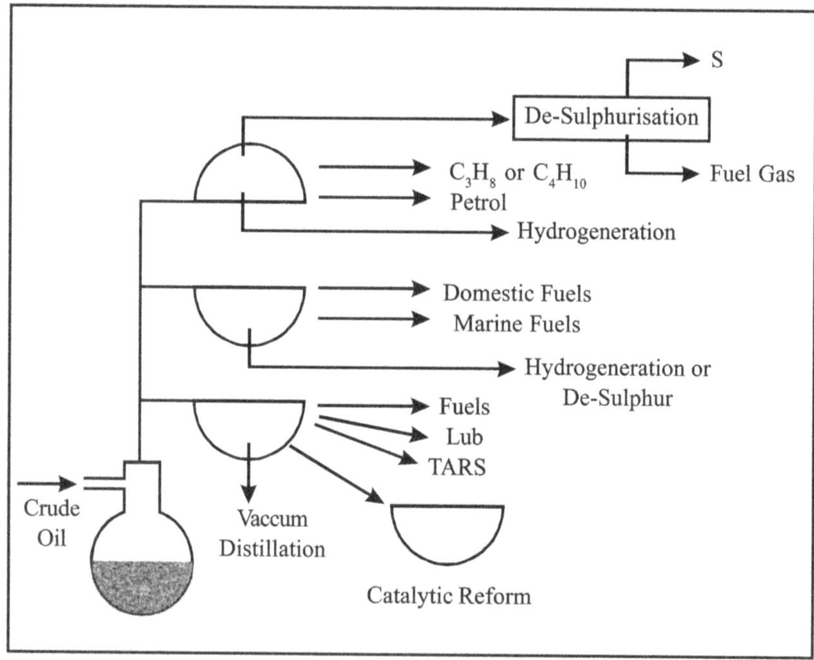

Fig. 4.2 Layout of a Petroleum Refinery

Wide-cut gas oils comprising of light lubes, heavy lubes and industrial oils distil out between 300 to 450 °C. The residual oils of the type bunker oil, asphalt etc. boil out at temperature above 450 °C. It is to be noted in this connection that the refineries are designed in accordance to the nature of the crude petroleum they receive for processing. A line sketch of the refinery is shown in Fig. 4.2. In Table 4.5 are shown the various fractions obtained on refining of crude oil at the refinery.

Table 4.5 *Fractional Distillation of Petroleum*

Fraction	Boiling Range°C	Fuel Utility
Gases - Ethane, Butane, Propane	Upto 50	Domestic fuel
Light gasoline	40-75	Blends of automobile fuels
Naphtha	75- 200	Fuel, solvent, synthesis
Kerosine	170- 220	Aircraft and domestic fuel
Diesel and gas oils	200-350	Automobile, engines, heating
Residue	Above 350	Heavy fuel oil, cracking etc.

The gases that distil out from the crude oil at the onset of the refining process are collected in huge tanks. They constitute mainly methane, propane and butane which find immense use as domestic and industrial fuels. The yield of this mixed gases is to the extent of 3-5% by weight of the crude oil. The individual gases need not be separated but collectively compressed and bottled. When this mixture is in compressed form it assumes a liquid state, and hence it is notified as 'liquefied petroleum gas' (LPG).

The light straight run gasolines are comprising of solvents, hexane, heptane, octane, nonane and a few iso-paraffins too. These are also fuels in motor cars (as petrol) and petrol-run-engines. The yield of this fraction is to the extent of about 22% by weight of the crude.

Kerosene which collects in the middle run distillates to an extent of about 13% is a fuel for aircraft and jet engines. It also finds use in domestic life as a fuel for lighting.

Diesel oils originate in the middle run distillates amount to approximately 26% and are very prominent fuels in the internal combustion engines and turbine engines .

Furnace fuel oils, several industrial oils, light lubes and heavy lubes are recovered to an extent of 9-12% from the wide-cut fraction in the refining process.

The residue that remains in the still is a source for a variety of materials such as waxes, asphalt mixtures etc. The yields of individual fractions during the refining process therefore depend largely on the type of crude oil taken. The properties of petroleum obtained from different sources are shown in Table 4.6.

Table 4.6 *Properties of Petroleum obtained from different sources*

Property	S-1	S-2	S-3	S-4	S-5
Colour & Appearance	---	---	---	---	---
Specific Gravity, 60/60°F	36.5	30.7	38.1	34.9	0.88g/cc, 20°C
Distillation Range°F	196-759	203-1047	178-994	155-1032	---
Viscosity, cs 100°F	37.3	64.3	35.7	41.7	18.9
Water Content %	0.11	trace	trace	trace	---
Metals- Vanadium, ppm	0.7	175	0.3	3	---
Nickel, ppm	0.9	17	5	12	---
Sulphur, %wt	0.16	1.48	0.14	0.56	2.84
Pour Point, °F	5	10	-35	10	-34
Flash Point °F	<40	<40	<40	<40	---
Vapour Pressure, mm/Hg	3.2	2.5	6.9	8.4	160

S-1 Texas Gulf Coast Mix; S-2 Venezula; S-3 Nigeria; S-4 Canada;
S-5 Arabian Crude
**Source: Encyclopedia of Chemical Technology, Vol 18, 1982/ASTM*

Based on the specific gravity, distillation residue and carbon content, the crude oils available from various sources in the world have been categorised in the following groups (Table 4.7).

Table 4.7 Types of Naturally Occuring Petroleum

Oil Type	Occurrence %
Paraffin- naphthenic	40
Paraffinic	19
Aromatic- naphthenic	7
Aromatic- intermediate	22
Aromatic- asphaltic	8
Naphthenic	4

CHARACTERISATION OF THE PETROLEUM FRACTIONS

All petroleum products are characterised from the data obtained on performing the following tests:

1. Physical state and colour
2. Specific gravity
3. Viscosity
4. Distillation range/ boiling range
5. Pour point
6. Flash point
7. Carbon residue
8. Ash content
9. Water content
10. Flame test
11. Unsaturation test
12. Smoke point
13. Sulphur and vanadium content
14. Acid value
15. Copper strip corrosion
16. UV- VIS spectra/ IR spectra
17. GC- MS data
18. TLC/ HPLC data

Standard methods recorded in Handbook of petroleum published by Institution of Petroleum, UK or American Petroleum Institute, USA or Indian Standards Institute, India are to be followed.

Cracking of Petroleum

It is observed from Table 4.5 that the total yield of the fractions that have direct fuel value turn out to about 68% by weight of the crude subjected for refining. The remaining high boiling residue fraction do not find much direct application, although a small quantity of lubes, waxes, tars and asphalt are recovered. Therefore, a process of splitting these residues into lower molecular hydrocarbons has been adapted. The objective of cracking of high boiling residue fractions of petroleum is to disintegrate their molecules to low molecular weight compounds, thereby obtaining some 'more utility' products. By this technique the value of the 'residue fraction' is enhanced with respect to the 'more utility' oil fractions it renders. The basic processes of cracking are:

Thermal Cracking

Here decomposition of the residue fraction is achieved by application of heat either in vacuum or with superheated steam. Low molecular alkenes such as ethylene and propylene are obtained.

Catalytic Reforming

In this process metal oxide catalysts, such as, V_2O_5, FeO, MoO, and ThO and heat are employed; aromatic compounds, namely benzene, toluene, several xylenes which are starting materials in synthesis are produced. The use of catalysts in the cracking minimise the requirement of high temperatures.

Steam Reforming

In this the residues are subjected to turbulent reactions with steam. Carbon monoxide and hydrogen, the basic energy chemicals are obtained. A line sketch of the petroleum cracking plant is drawn in Fig. 4.3.

The basic chemical reactions that take place during 'cracking' have been written hereunder:

$$Naphtha \rightarrow CH_4 + C_2H_6 + C_3H_8 \text{ and so on}$$

$$CH_4 \rightarrow CH_2 + H_2$$

$$CH_2 + CH_4 \rightarrow C_2H_6$$

$$C_2H_6 \rightarrow C_2H_4 + H_2$$
$$C_2H_4 \rightarrow C_2H_2 + H_2$$
$$C_2H_2 \rightarrow 2C + H_2$$
$$CH_4 \rightarrow HC = CH$$
$$CH_4 \rightarrow C_2H_6$$

The naphtha, kerosine and several uneconomical sources of oil such as 'tar sands, oil shale' are successfully subjected to 'cracking' operation to afford a variety of synthetic fuels (Fig. 4.3). The calorific value of many of these synthetics are quite high and therefore they are also energy resources.

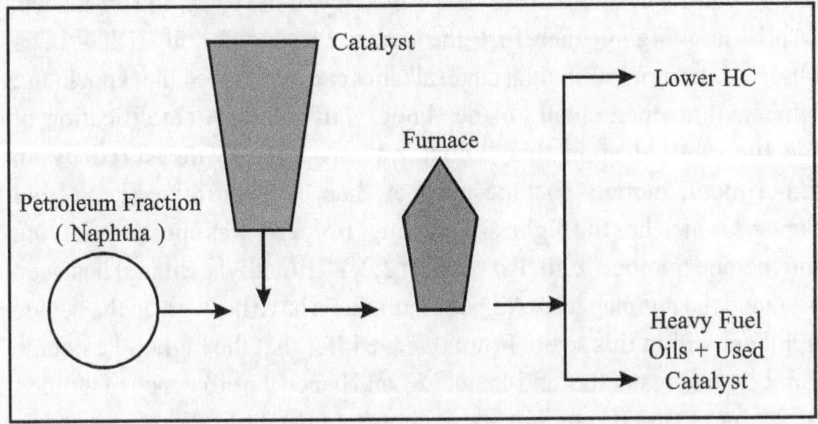

Fig. 4.3 Line Drawing of a Petroleum Cracking Unit

ANTI-KNOCKING IN IC ENGINES

In the IC engine the cylinder is the unit where ignition of the gasoline and air mixture takes place. The piston of the cylinder is in down position when the fuel mixture is slowly admitted in the cylinder by the carburettor. Later, the piston is moved to the up position when the fuel mixture is subjected to compression. The ratio of the volume of the space occupied by the fuel mixture when the piston was in down position to that of the volume when in up position is called 'compression ratio'. The efficiency of an engine is judged by the high compression ratio.

When the piston experiences the up stroke, that is, high compression, a spark is made by the ignition system in the cylinder, thereby burning the gases and expanding them and pushing the piston again down. All this mechanism takes place under ideal situations. However, when the compression ratio exceeds the limit, the ignition of the fuel takes place before the piston reaches the end of its up stroke. As a result of the early combustion, the gas mixture at the end causes an explosive type noise, termed as 'knocking'. Also, the ignition temperature of the fuel mixture is lowered. In fact, when ignition takes place at the right moment, the power generated by the fuel would be fully utilised for driving the engine; otherwise, on account of the knocking phenomenon much of the energy was lost. Automatically the efficiency of the IC engine is reduced. 'Anti-knocking' is therefore the desired phenomenon. It has been observed that normal straight chain alkanes cause considerable 'knocking', whereas branched chain alkanes knock fairly less. A classification of the fuels based upon this 'knocking' property as measured by an experiment, namely 'octane number' has been introduced. Normal heptane which has the highest 'knocking' property was chosen as having the 'octane number' zero. Iso octane (2,2,4- trimethyl pentane) has been assigned the number- 100. All petroleum fuels have been given the octane numbers within this scale. From this we infer that the higher the octane number, the greater the 'anti-knocking' and hence the efficiency of ignition of the fuel in the IC engine. Some important generalisations were made in this connection as regards the structure- performance of the hydrocarbon fuels. These are :

- In the alkane series, the octane number decreases with the increase of the carbon chain and increases with the branching.

- In the homologous series, all alkenes have higher ratings of octane numbers than their corresponding alkanes.

- In a straight carbon chain hydrocarbon structure, if the position of the $C = C$ double bond is shifted to the centre, the octane number is higher.

- The olefins, cycloparaffins and aromatic hydrocarbons possess high octane numbers than their n-paraffin counterparts.

- The octane number of gasoline used in India is between 75-80.

In the past, organo-metallic compounds of the type lead tetraethyl were added to the petroleum fuels in order to improve the octane number. Since the use of lead is banned nowadays on account of its detrimental effects on the air environment, several branched hydrocarbons, naphthenes and aromatics are blended with petrol in place of lead tetraethyl. These organic blends are also not free from causing pollution and so the IC engines are fitted with catalytic converters to effect complete combustion of the fuel including the additives.

The combustion of fuel in a diesel engine is different. In this, the air alone is compressed together with heating to about 290-340 $^\circ$C. At the end of the compression stroke the diesel fuel is injected into the system and instantaneous ignition occurs causing the piston to drive off the engine. The diesel fuel need not be as volatile as petrol; also, no electric spark is required to ignite the fuel. The quality of diesel is expressed in terms of 'cetane number'. For expressing the cetane number, a mixture containing cetane, (n-hexadecane), $C_{16} H_{31}$, value = 100 and -methylnaphthalene, value = 0, has been taken and all fuels assessed against this mixture. Usual diesel engines require a fuel of cetane number above 45 for nice performance.

Automobile Emissions

All automobiles which use petroleum fractions as fuel for performing work emit three pollutants into the atmosphere. These are carbon monoxide, hydrocarbons and nitrogen oxides. The quantity of each of the pollutant differ from the type of the engine operation e.g., idling, cruising, acceleration and deceleration. In the case of petrol engines, the chief emissions are CO. HC respectively during all these operations showing much higher values than a corresponding diesel engine. In diesel engines however, the emissions of NO_x are much higher than the similar petrol engine. Also, diesel engines leave a lot of soot while operating. In Table 4.8 are shown the comparative data.

Table 4.8 *Automobile Emissions during Operation of Petrol & Diesel Engines*

Pollutant	Idling		Cruising		Acceleration		Deceleration	
	D	P	D	P	D	P	D	P
CO	Traces	52000	20	8000	500	42000	Traces	52000
HC	250	750	65	3000	115	400	250	4000
NO_x	60	30	1500	850	850	300	10-50	12-30

D = Diesel Engine; P = Petrol Engine; All values are in ppm.
Source: *Y. Anjaneyulu, Text Book of Air Pollution and Control Technologies, Allied Publishers, 2002, p494.*

NATURAL GAS

Natural gas is invariably present along with oil in the wells and is brought to the ground in the dissolved state. In some instances it forms a cap over the oil and drives out first from the wells. Some wells exclusively deliver natural gas. Natural gas is explored and brought out over the land in the same manner as the crude oil from the wells. The gas coming out is usually at a pressure of 400-500 bar and therefore must be allowed to expand on release from the well. Later the components such as water, nitrogen and hydrogen sulphide are systematically removed from this gas to give a pure form.

North America and Middle East countries have huge reserves of the gas. The major constituent of natural gas is methane. If the methane content of the gas is 95%, it is termed as 'dry natural gas' and if significant quantities of ethane, propane and butane are also accompanying methane, it is termed as 'wet natural gas'. The composition of most commonly occurring natural gas is: methane 80%, nitrogen 5%, ethane 10%, propane 4%, butane and others 1% respectively. Hydrogen sulphide is a contaminant in the natural gas. Even helium gas is found along with natural gas of American origin. Both helium and sulphur are recoverable from natural gas easily. For the purposes of transporting the gas by sea the natural gas is subjected to liquefaction in a plant to afford LPG. The LPG occupies nearly one sixth the volume for the same weight of the natural gas.

Since natural gas is rich in hydrocarbons it can be used as a fuel. India's needs of energy are met to an extent of 7% by natural gas. The gas is produced from the wells situated at Bombay High, Heera, Panna, South Bassein, Neelam of the west coast and the Krishna - Godavari- Kaveri basins of eastern side. Andaman- Nicobar islands are also showing promising holdings of this gas. The 1999 statistics put the figures of natural gas production and consumption at 0.750 and 0.752 trillion cubic feet (tcf) respectively. It is expected that with rapid industrialisation both these figures will go up and a good future is in the offing for gas- based power plants as well as petrochemical production.

Bottled natural gas is a liquid and is known as liquefied petroleum gas (LPG). Compressed natural gas (CNG) is currently used as an automobile fuel in India. Domestic cooking fuel lines were laid in many coastal cities such as Mumbai and Mangalore in the Konkan coasts. It is a clean fuel compared to the other fossil fuels, as the sulphur pollution is minimal and the CO contribution into the atmosphere is also low. Even thermal power plants utilising natural gas are functioning and the one in Trombay near Mumbai in India is famous. Natural gas is an economic fuel in industry too. The gas is used for space heating in many European countries. Many petrochemicals, pesticides and fertilisers are manufactured from this material.

Standards for Natural Gas

The ASTM and American Institute of Petroleum or British Standards are commonly adapted to test the quality of natural gas. The average calorific value of house supply natural gas is mentioned as $0.04\ GJ/m^3$.

TAR SANDS AND OIL SHALE

Tar sands are embedded with viscous bituminous oils that cannot be easily recovered in the drilling of earth. The countries, namely Venezuela, Canada and Russia have these reserves. The isolation of oil from the sands is a tedious operation which involves the use of a lot of land, water and power. Therefore they were not fully exploited.

Oil shale is widely distributed in the states of Colorado and Utah of USA and in Brazil. In north-east India some rocky deposits of shale are found and it is estimated that about 14 billion tons of oil can be obtained from these rocks. Almost one ton of oil shale is required to produce one barrel of oil. Moreover, the waste obtained in the extraction is equal to that of shale taken. Disposal of the waste is another problem. These are some of the constraints in establishing the machinary for extracting the oil.

Both tar sands and oil shale when subjected to thermal cracking yield synthetic gas which is easier to collect and utilise for power generation.

SYNTHETIC FUELS

Synthetic fuels, also known as Synfuels are liquid or gaseous mixtures of low-boiling hydrocarbons which have good combustion characteristics. They are derived from coal, coal tar, specific petroleum fractions, tar sands, oil shale through hydrogenation process or subjecting to steam heating, and from wood and biomass through destructive distillation process. Typical synfuels are the coal gas, producer gas, carburetted water gas, coke gas and wood gas. The chief objective is to increase the hydrogen level thereby bringing a low C/H ratio. Alternatively the decrease of carbons in the material through pyrolytis also brings in certain synfuels which possess good combustion characteristics.

Synthetic fuels include low, medium and high calorific value gases, liquids and clean solids. Producer gas obtained from coal with calorific value 3.5-10 MJ/cu.meter is a valuable fuel for generation of power in turbines. Medium calorific synfuels possess CV in the range of 10-20 MJ/cu.meter and high type about 35-38 MJ/cu.meter. In fact during the coal gasification process using steam and air, followed by purification on removal of H_2S, NH_3, CO_2 , a variety of these low, medium and high CV gases are secured. The synthetic natural gas (SNG) thus obtained is a clean and superb synfuel in industry.

The simplest route for obtaining synfuels from oil shale and tar sands is to conduct pyrolysis and treat the resultant liquid mass with steam, remove the undesirable H_2S and NH_3 and distil the resultant liquid. A typical scheme is shown below:

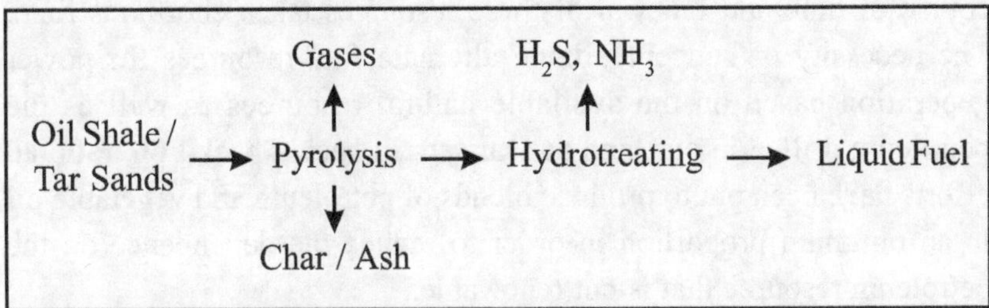

The C/H ratios for a few fuels is shown below:

Fuel	Molecular Formula	C/H ratio
Bituminous Coal	$CH_{0.8}$	~15
Benzene	CH_1	12
Crude Oil	$CH_{1.33}$	9
Gasoline	CH_2	6
Methane	CH_4	3

The organic material in both tar sands and oil shale has a C-H mass ratio of ~8 which is close to that of crude oil and for this reason to process these materials for obtaining synfuels is easier than that of coal. A typical assay of pyrolysate of oil shale is as follows:

Gas	Volume per cent
CO	5.5
CO_2	36.5
H_2S	3.2
H_2	18.7
HC	35.6

Pyrolysis at 500°C and Above

Biofuels and Biodiesel

The issue of energy security assumed great importance at the present juncture in most countries that have shortfall in the production of fossil fuels. As well known, the availability of fossil fuels diminish in course of time and renewal of these resources takes enormous time. The necessity of conceptualizing alternate fuel resources for power generation based on the available natural resources as well as the petroleum drilled in our land has arisen. In such a backdrop, rational efforts have been put to produce blends of petroleum and vegetable oil in an optimum proportion in order to reduce the dependence on the petroleum resource that is not renewable.

It is fascinating to know in this connection that Henry Ford during the first decade of nineteenth century ,designed an automobile engine for T-model that was expected to run on ethanol alone. In the year 1895, Rudolf Diesel designed the diesel engine with the idea of running the same on many fuels, including peanut oil. He displayed the engine in the "world exposition" in 1900 and operated it using peanut oil. The establishment of the oil firm namely Standard Oil Company in 1870 by John D. Rockfeller revolutionized the functioning of diesel engines throughout the world with the usage of diesel oil and this practice remained till the present times; however in the present times, due to the rising demand for fuel and the heavy budget towards its procurement, the necessity of an alternate fuel, namely biodiesel has arisen. The various techniques of its manufacture, analysis, application/substitution, toxicity testing, standardization, consumer safety practice and transport and storage have been described in the following pages.

Two indigenous organic materials of plant origin that mix up easily with petroleum fuels and promise the same fuel value as petroleum are the ethanol, and the neat soybean or rapeseed or cotton seed or oils from non-edible seeds of jatropha (ratanjyot), and pongamia (karanja) and the like. The jatropha seeds that produce oil are non-edible and grow on degraded and forest lands and are most suited for oil extraction.

An intimate mixture (upto 10% of ethanol) of ethanol with petrol is a biofuel. The substitution of ethanol upto 10% in the petroleum oil do not significantly change the fuel characteristics at all and therefore offers a saving of the petroleum. The techniques of obtaining ethanol from various cellulosic and lingo-cellulosic crop materials using improved yeasts has been described in Chapter 10 of this book.

An optimum mixture of petroleum diesel with the vegetable oil, usually 20%, (that has undergone trans-esterification with methanol/ethanol) forms a biodiesel. Blending with vegetable oil do not warrant a change in the diesel engine design, nor alter the burning characteristics.

According to a US Department of Energy Datasheet, 2000, India ranks sixth in the world in terms of usage of primary energy. The country consumes about 9 million tons of petrol and 42 million tons of diesel annually and the requirement may go up in the coming years. The country imports crude oil from several oil exporting countries and the bill goes anywhere between 1.1 to 1.5 billion rupees annually. A blend of ethanol to petrol (reasonably to an extent of 10%), and biodiesel to petroleum diesel (to an optimum of 20%) would considerably reduce the import burden on country as well as generate more employment and industrial development. It is in this context that the National Mission on Biodiesel has been entrusted by the Government of India to undertake projects for enhancing the yields and using the same in blends of petroleum diesel in order to tide over the scarcity situation in diesel supply.

The manufacture of biodiesel involves the participation of agro-industrial sector, finance sector and policy makers. Starting from cultivation of quick yielding jatropha /rapeseed plantations, extraction from seeds, conversion to methyl esters through trans-esterification, i.e., a modification of the oil to a low viscosity, blending in mixers and filtration of the requisite blend, high skills are necessary and initial encouragement is required for introduction in the transportation network as well as farming sector.

Biodiesel production do not present many technical problems.

1. The plants chosen are usually quick yielding and grow on degraded and barren lands that are not costly.

2. The oil-yielding seeds are available in all seasons and do not require special care as regards planting or collection of produce.

3. The rural population secure jobs during the plantation time as well as at the stage of oil-expelling and blending operations.

4. The plantations render sustainable income, employment and soil moisture-preservation.

5. The oil cake left behind can be utilized for bio-gas generation, or as organic manure.

6. In India, eventhough the biodiesel is produced from non-edible oils, the process costs are profoundly higher and hence initial subsidies from the governments are necessary to establish the newly developed industries.

Biodiesel –Physical Characteristics

Colour :	usually light yellow
Specific gravity:	0.87-0.89
Kinematic viscosity at 40 $^{\circ}$C	3.7-5.8
Cetane number:	46-70
Sulphur, wt%:	0-0.002
Cloud point,	C : -11 - 16
Pour point,	C : -15 - 13
Iodine number:	60-135
Lower heating value:	15,700-16,735
Higher heating value:	16,928-17,996

Biodiesel oils age faster than the fossil diesel on account of the presence of fatty acid methyl esters present in it. The oxidation stability of the blends is also poor. Further, the storage of neat biodiesel (B 100) or the blends (such as B20) in a tank of a vehicle beyond 6 months is not recommended.

Emissions in Engines

All biodiesel blends have been found to be compatible with the existing diesel engines. The emission of sulphur dioxide from these blends has been observed to be lower than that arising from petroleum diesel. However the Nox emission in the case of biodiesel is found to be higher. The following table shows a comparison in a test run.

Table 4.9 *Tests result for DDC series 50 engine at transient conditions*

Test fuel	HC (g/hp-hr)	CO (g/hp-hr)	NO_x (g/hp-hr)	PM (g/hp-hr)
Diesel	0.06	1.49	4.5	0.102
B20	0.06	1.38	4.66	0.088
B100	0.01	0.92	5.01	0.052
B100 with catalyst	0.02	0.76	4.9	0.03

Compatibility with Fuel Additives

It has been observed that the common diesel fuel additives do not serve in the operations involving blend biodiesel; hence R&D efforts are to be put to select suitable additives for biodiesel blends.

Standards

Pure biodiesel(B100) and blend stocks: ASTM D6751.

Personal protection : Safety glasses & adequate ventilation

Transport care : Hazardous on air, sea and road freight; stored in 5 and 30 kg containers.

Applications : Biodiesel is widely used in stationary pumpsets and other agricultural engines. It is also widely used in diesel cars, marine engines and sea boat engines.

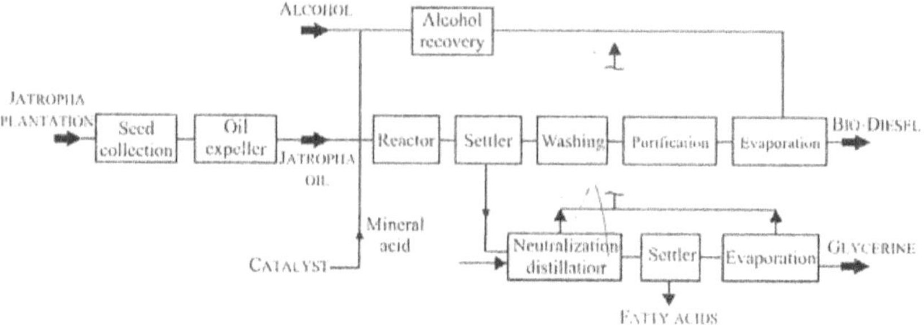

Fig. 4.4 Production of bio diesel using jatropha as feedstock to be given.

The flow sheet for production of biodiesel starting from jatropha feedstock is shown below:

ENVIRONMENTAL IMPACTS

Many operations during the mining of coal are associated with the damage of the environment. Invariably carbon monoxide and particulate emissions occur in the mines and surrounding townships. Regular monitoring of these pollutants help in checking the environmental damage. Even the water that is used in the drilling operations carry a few inorganic substances such as vanadium, cadmium, sulphur, strontium, uranium and lead and wet the surrounding areas causing soil pollution. The conversion of coal to coke, the pre-washing of coal for removal of sulphur, the destructive distillation process of coal tar, deposition of pitch and several operations leave pollutants into the atmosphere. Finally, the burning of coal in the thermal power plants generate enormous quantities of sulphur and nitrogen oxides that produce acid rain. The flyash that accumulate at the thermal power plants pose a serious environmental problem, unless rapidly cleared by ancillary brick making industries.

Similar problems as were faced during the mining of coal were observed in the drilling of petroleum from the earth. Here, the evolution of natural gas, a component of petroleum gusts into the atmosphere causing a bit of air pollution in the zone. The mud waters that flood the wells also bring inorganic salts on the upper ground thereby causing soil pollution. At the refinery, emissions of sulphur dioxide do occur and this has to be checked properly. The 'flare' which is part of a refinery emit left over gases most of which are carbon oxides, and these contribute to global warming. Besides, the mercaptans, nitrogenous bases, peroxy acetyl nitrate, ozone and several aldehyde substances are formed in the atmospheric photochemical reactions resulting from the petroleum hydrocarbons and atmospheric nitric oxide. The emissions of sulphur and nitrogen oxides in the atmosphere coming from burning of fuels in the power plants is a common feature. Similar environmental problems are faced during the use of natural gas, shale oil and other synthetic fuels derived from the fossil fuels. It is obvious that if the sulphur in the fossil fuel is isolated by any of the conventional techniques, subsequent environmental pollution is minimal. Moreover, the isolated sulphur is another energy resource, as it is the raw material in the production of sulphuric acid. Even cadmium, silver, vanadium, molybdenum can also be recovered from the fossil fuels before subjecting them for processing.

Questions

1. Describe the formation of coal beneath the earth. State the different varieties of coal available in the world and elucidate their properties.

2. What is destructive distillation of coal? Describe the set of conditions under which this is performed and characterisation of the products.

3. How is coal tar distillation performed in refinery and which are the products obtained?

4. What is water gas? Give its composition and calorific value.

5. Write the composition of producer gas and the heat value.

6. Briefly explain the formation of petroleum in earth and its drilling and refining operations. Which particular constituents of crude oil are used for 'finger printing' the origin of the crude?

7. Describe the cracking of petroleum fractions. Explain how this process render more utility products.

8. Describe 'anti-knocking' in an IC engine.

9. Write notes on octane number of petroleum fuels.

10.   Discuss the automobile emissions in engines during various operations.

11.   What is 'dry natural gas'? how is it secured from the earth bed? What is the utility of natural gas?

12.   Give an account of synthetic fuels, discuss the utility of these fuels in industry.

13.   Explain the environmental impacts caused during the mining and usage of coal and its derivatives. What is fly ash and how is it useful in industry?

14.   Explain the effects of over-utilisation of fossil fuels in our country. Discuss the usage and adaptation of an alternate energy resource for regular power generation.

15.   What are 'biofuels'? How are they different from the fossil fuels?

16.   Discuss the scope of blending biofuels with the regular petroleum fuels while maintaining the calorific requirements.

17.   Present a detailed account on Prospective cultivation of Jatropha plants in India to render biofuels.

5

SOLAR ENERGY

The Sun moving at a distance of about 93,000,000 miles from the Earth is the source of energy on earth. Only a fraction of the energy transmitted by the sun is ultimately absorbed on earth and that transforms into different types of energy. The total solar power output in space is reckoned as 1.2×10^{36} J/year. Out of this energy, 5.4×10^{24} J/year is intercepted by the earth. About 65% of this intercepted energy, namely 3.5×10^{24} J/year usually gets absorbed in the atmosphere and on the earth's surface. About 10% of this absorbed solar power (3×10^{23} J/year) turns into wind power that later on dissipates into heat, ocean waves. Ultimately a very small percentage of incident solar radiation is utilised to bring in photosynthesis in plants on earth. The energy received on earth from sun is termed 'solar energy' and it reaches in the form of radiation. It was established that solar radiation on earth produce the energy to the extent of about 1300 W per sq.meter, that is 1 calorie per square centimeter.

Solar Radiation

The value of mean distance between the sun and earth is 1353 W/m². The earth receives energy from sun in the form of electromagnetic radiation i.e., visiable light, infrared, ultraviolet, X-rays and radiation waves. Electromagnetic radiation travels through space at the speed of light (3.00×10^8 m/s) and each type of radiation is classified by a frequency, and a wavelength. The most intense solar radiation is in the wavelength region of visible light. Intensity is reduced for any wavelength in the infrared region and is very weak in the ultraviolet region. The spectrum at the earth's surface is reduced from its intensity at the top of the atmosphere, because of the absorption and scattering properties of the atmospheric gases (Fig. 5.1).

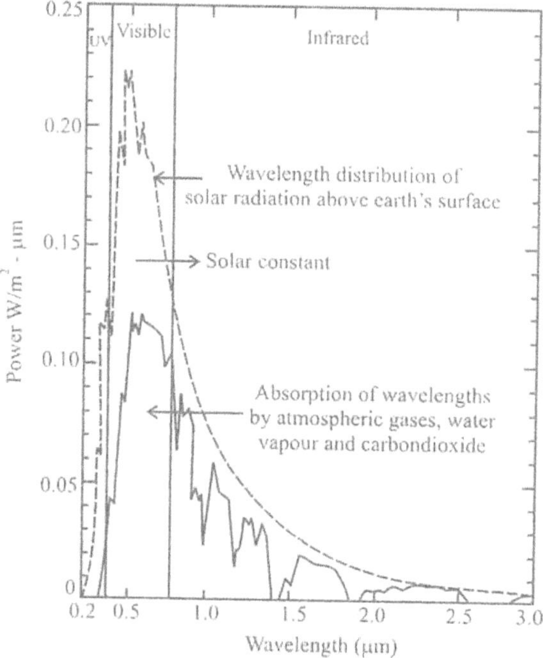

Fig. 5.1 Absorption of wavelengths by atmospheric gases.

The Earth makes a complete rotation around the sun once a year. The earth itself rotates on its north-south axis once every 24 hours, which accounts for one night and one day. In the summer the Northern Hemisphere is tipped towards sun, so the earth receives sun light more directly and in the winter the earth tips away from sun so that, the sun light comes to earth at a lower angle. This accounts for longer warm days of summer and shorter colder days of winter. At the North Pole the sun is in the sky 24 hours a day for 6 months centered around June middle but the sun is not visible for another six months of the year thereafter. The variation in solar intensity over day and night, summer and winter, as well as changing cloud cover complicate the use of solar energy (Fig. 5.2). Therefore, it is essential to store solar energy to utilize at the time of reduced sunlight. The solar radiation thus does not all reaches on the upper atmosphere of the earth, because it is partially absorbed and scattered by the atmospheric gases and the clouds.

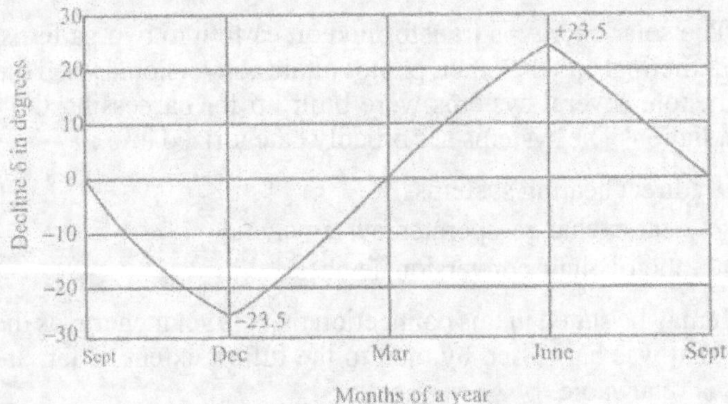

Fig. 5.2 Variations of Sun's decline on earth.

Incident Light

The light rays emitted from sun travel through the air medium towards the surface of earth. The rays to start with take a straight line path but the atmospheric dust, water vapour and some objects which are situated in the path cause scattering of the rays. Therefore, the light incident on earth comprise of mixed parallel beam(direct light) and also scattered light which do not possess parallel rays. However the solar constant is determined from a total accumulation of the energy generated from both types of incident light. The solar energy collectors also receive both parallel beam radiation and diffused/scattered radiation.

Solar Insolation

The solar radiation received on a flat horizontal surface at a specific location on earth at a specific time is termed as 'solar insolation' and this is expressed in units:

$$SI = Watts / sq. Meter$$

The solar insolation is determined by the following parameters:

1. Geographical location, eg. latitude
2. Area of surface, sq. Meters
3. Hourly and seasonal climate variation
4. Atmospheric clarity
5. Angle of tilt of solar incident radiation
6. Interferences by tall structures, trees and adjacent solar panels

Based on this data 'solar insolation map' for different Indian cities have been developed for rapid reference.

The solar energy is transformed on earth into two patterns: one, the solar thermal and the other, photovoltaic conversion. Based on these phenomenon, several systems were built up for harnessing the entire solar radiation. The systems are broadly categorized into :

- direct heating systems,
- heat exchange operated systems, and
- photovoltaic conversion devices.

It may be stated in this connection that the solar energy is the only energy that was harnessed by man to the fullest extent so far, and this energy is renewable.

DIRECT HEATING SYSTEMS

The following are some of the household gadgets that have been developed making use of solar radiation incident on land. (see Fig. 5.3). No extra effort is put to collect the radiation except in the most natural form.

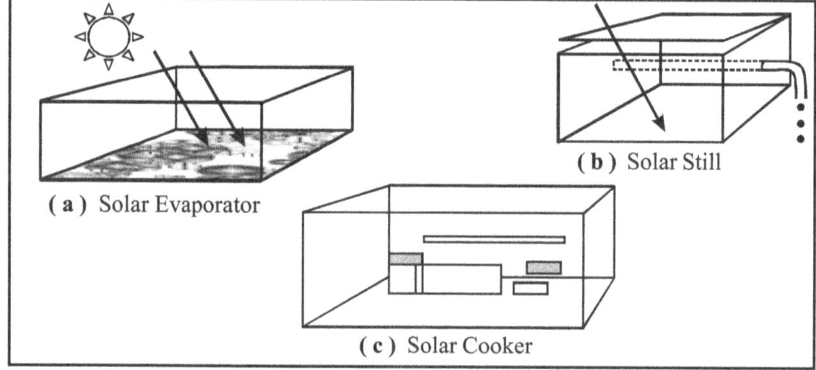

Fig. 5.3 Solar Direct Heating Gadgets.

Solar Evaporator

A stainless steel tray of dimensions $18 \times 12 \times 4$ inches that holds fluids requiring concentration is the vessel. The vessel is placed at a location where sunlight is falling directly, and in course of time, the solvent in the fluid gradually evaporates giving a 'concentrate'. Besides household applications, solar evaporators find use in food processing, common salt manufacturing and chemical industry.

Solar Heater

An aluminum chest $18 \times 12 \times 9$ inches size with separators that stand directly exposed to sunlight is a solar cooker. The separators are charged with the raw foods and requisite water and placed in the open sun, when the materials absorb the radiation, get heated and give cooked food. A sunlight focusser arranged on the chest cover is sometimes focussed to allow rapid heating to take place.

Solar Water Heater (Thermo Siphon)

A free standing thermo siphon water heater is a household utility in cold regions. The sunlight falling on metallic plates produces heat and this energy is in turn absorbed by the circulating water which gets collected through siphon effect in a free-standing water tank.

Solar Distillation Unit

Solar energy in the form of direct sunlight is employed to effect distillation of water that is locally available to afford pure distilled water that is potable. This unit is particularly useful in desert regions, marooned ships, military stations and mountain areas.

Space Heating in Buildings

All south faced buildings with glass windows allow solar radiation to penetrate the interior spaces and absorb the heat. Even the air inside the building gets heated. This type of 'direct gain' of energy was welcomed since the medievial ages in Europe. If the building is thermally insulative, the heat is retained day and night and offers an energy saving upto about 20%. Currently architects are considering these factors while designing new structures in order to save energy spent for space heating of interior building.

HEAT EXCHANGE OPERATED SYSTEMS

In these systems solar energy is first absorbed on a collector and the heat is exchanged with another working fluid. Depending on the solar thermal applications the choice of the collector and working fluid are made. The mechanism of heat exchange can be explained from the three Rankine cycles that these systems follow. These are described in the line diagrams depicted in Fig. 5.4. In Fig. 5.5 are shown a few solar radiation collectors.

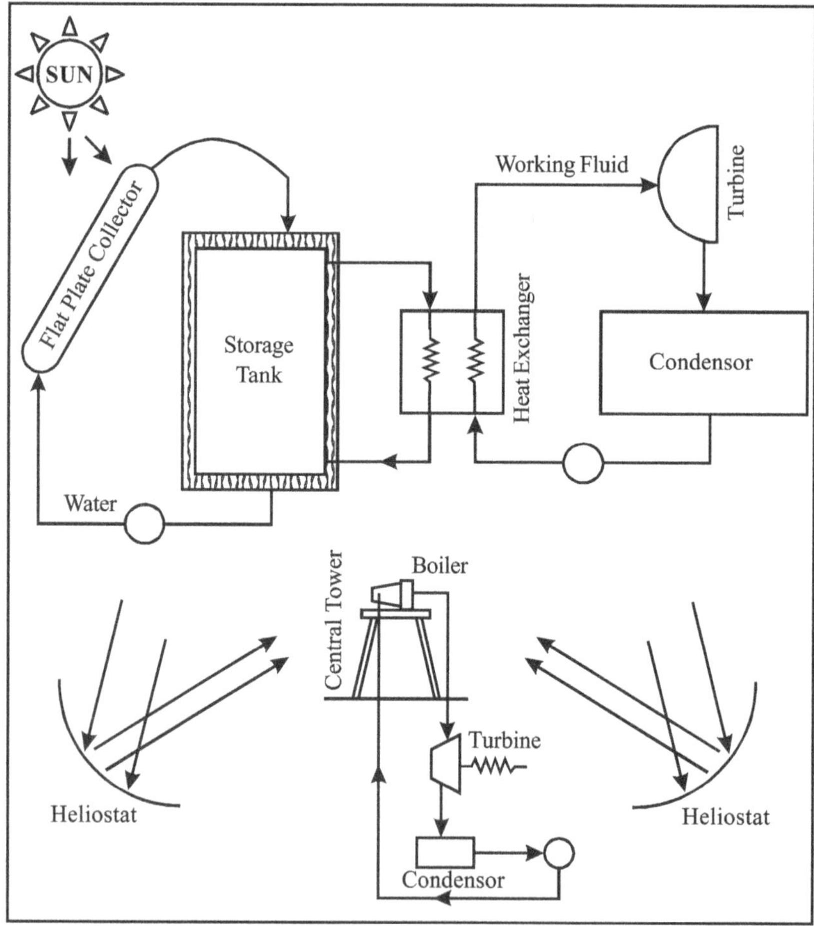

Fig. 5.4 Rankine Cycles-Low & Medium Types

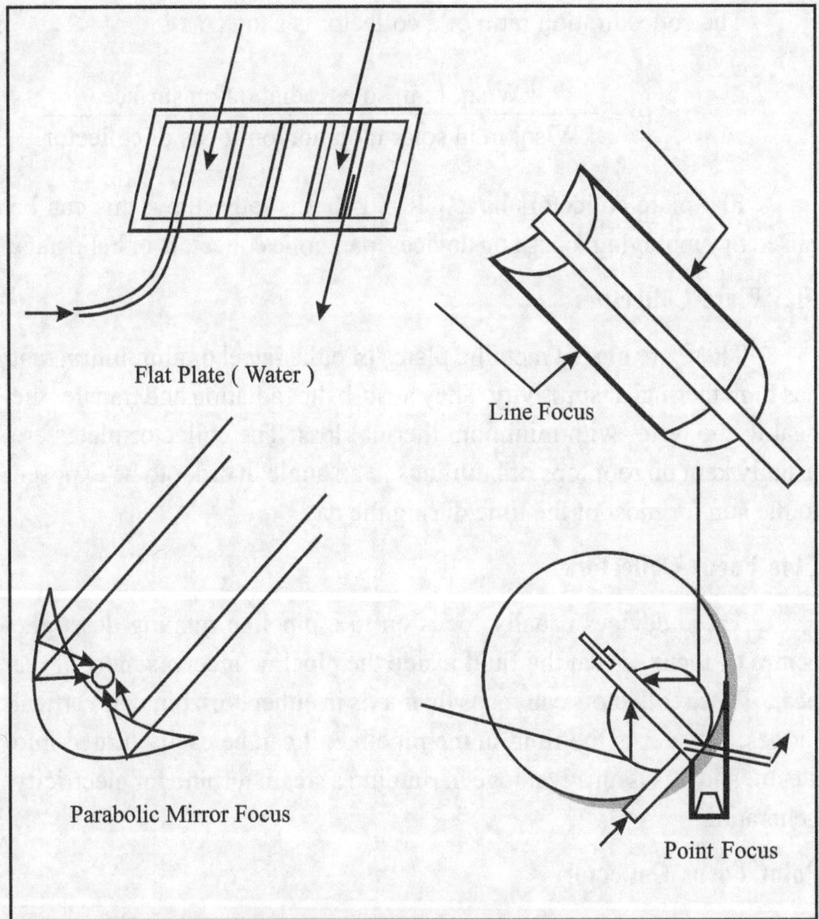

Fig. 5.5 Solar Radiation Collectors

Solar Collectors

The solar energy collectors have several designs to suit the requirement. Basically the collector must have large surface area and must be capable of transferring the heat to fluids that surround its surface. Sun light as such has low power density, namely 0.1 - 1.0 kW/sq.meter and hence the collector surface is expected to be very large to tap significant energy. The collector efficiency, lay out, tracking and atmospheric clarity play important role.

The concentration ratio of a collector is shown as :

$$CR = \frac{kW/sq.\ m\ in\ solar\ radiation\ on\ surface}{kW/sq.\ m\ in\ solar\ radiation\ on\ focus\ of\ collector}$$

Flat plate collectors have CR = 1 that is rather low; this can be raised by employing focussing devices, parabolic collectors or heliostats.

Flat Plate Collectors

These are glazed metallic plates of either steel or aluminium that has high thermal absorptivity. They absorb the radiation and transfer the heat to the water with minimum thermal loss. The collector plates are usually kept on roof tops of buildings in an angle in order to be exposed to the sun for most of the time during the day.

Line Focus Collectors

These devices usually focus on to a pipeline running down the centre of focus so that the fluid inside the pipeline receives most of the heat. These collectors can focus their axis in either horizontal or vertical modes. If water is the fluid in the pipeline, it can be easily turned into steam; and this is an advantage in running a steam turbine for electricity generation.

Point Focus Collector

These collectors track the sun and effectively absorb the radiation. They are quite useful for generating steam from a running water pipeline that is in contact. Even diffused sunlight is also focussed over the pipes.

The working fluid that is in contact with the collector is commonly water. If the operating temperatures of the system is below 100 °C this fluid can be run in the device; if it is still lower, say 50-60 °C the heat from the hot water can be exchanged by low boiling liquids such as freons. If the operation requires a temperature more than 100 °C the working fluids selected shall be propylene glycol and mixtures.

Solar Thermal Engine

The first solar thermal engine operating on steam was demonstrated by Frank Shuman in the year 1914. This engine produced work to the extent of 5.5 horsepower. The sunlight was focussed by parabolic trough collectors for obtaining the steam (Fig. 5.6).

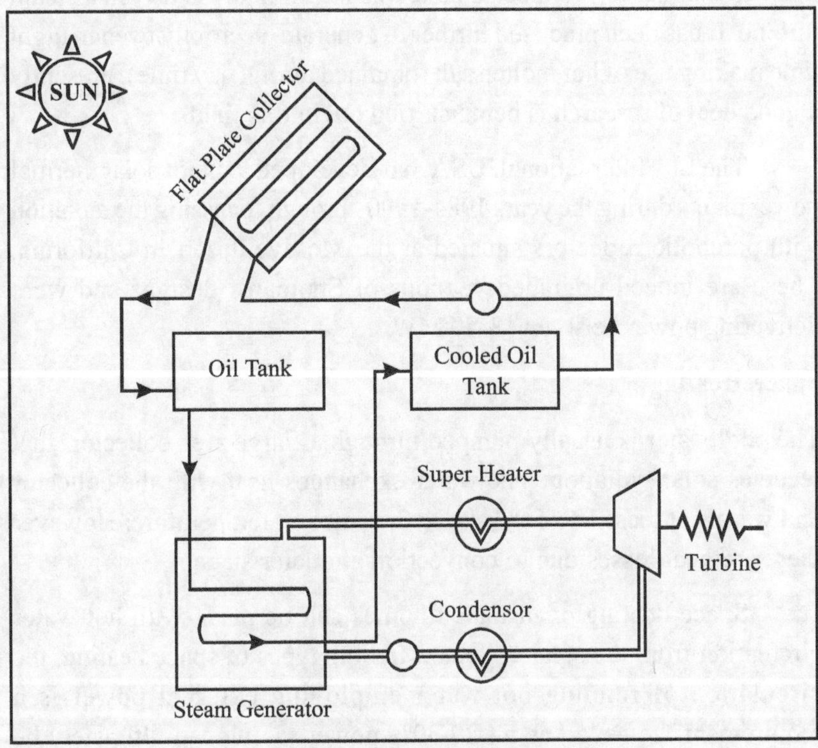

Fig. 5.6 *Solar Thermal Engine*

In a typical solar thermal power plant shown in Fig. 5.6 the steam obtained through the absorption of solar radiation by the circulating water in the parabolic trough collector is led into a steam turbine. In this the mechanical energy produced by the steam is converted to electricity and the resulting condensate water flows back into the hot water-steam boiler and finally goes into the parabolic trough collector water circulating system.

Making use of the 'field tracking heliostats' which reflect the sun's rays on to a boiler at the top of a central tower, a workable 10 MW solar thermal power station was constructed at Barstow, California, USA. The solar radiation that concentrated in the tower heats up molten rock salt/ or synthetic oil. This profound heat was later exchanged with water in an exchanger to produce superheated steam that has driven a steam turbine. It has been practised further to generate electricity even in night time making use of hot molten salt (obtained during day time). Presently a good deal of research is being carried out in this field.

The Luz International, USA, has developed 9 major solar thermal power plants during the years 1984-1990 through focussing the radiation with parabolic reflectors situated at the Mosave desert in California. These are indeed upgraded versions of Shuman's designs and were delivering power between 13-80 MW.

Space Heating

The pool water is usually pumped through a 'large area collector' that receives solar radiation. The water exchanges heat with the collector and warms at least 10-15 °C above the ambient temperature. However there are heat losses due to convection at a later stage.

Space heating of insulated homes can be made with hot water circulation from the solar collector. In both types of space heating, the circulation of running hot water employing external power is a requirement; however, about 20-30% power saving can ultimately be achieved on audit.

Solar Heat Pump

The design of a solar heat pump is akin to the Rankine cycle (medium) wherein the heat is exchanged by a second fluid that moves through a space and exchanging the heat. Side by side the cooling occurs in the chamber from where the heat is taken by this fluid to once again become a liquid. The whole process is depicted in Fig. 5.7.

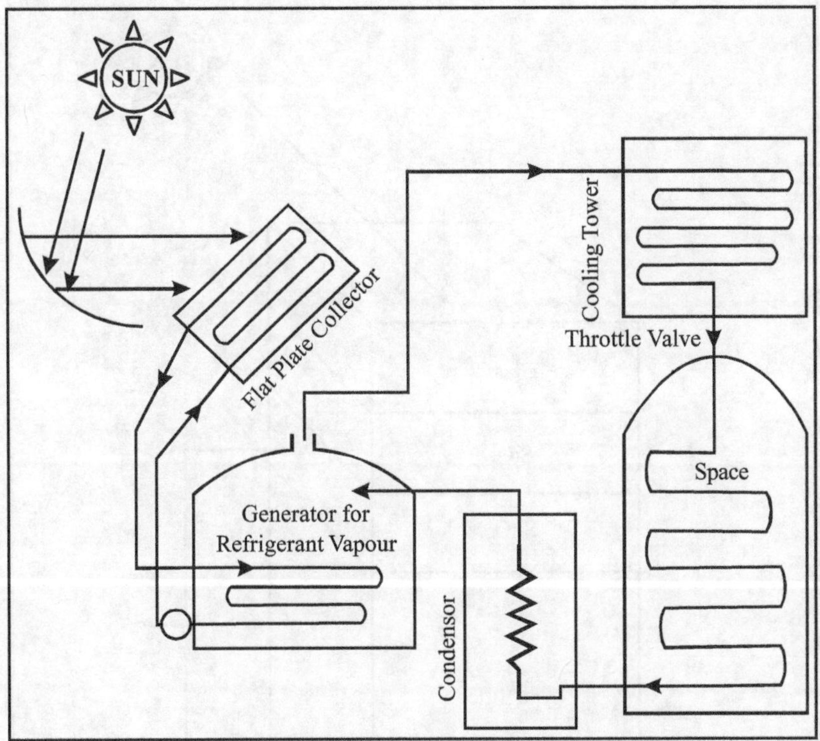

Fig. 5.7 Solar Heat Pump / Refrigerator

PHOTOVOLTAICS

Photovoltaics are solid state devices in which the sun's light is directly converted into electricity. The discovery of photovoltaic effect dates back to 1877, when two scientists namely Adams and Day, in Cambridge, observed the electricity generated by a film of selenium exposed to sunlight. By the year 1950, the Bell Telephone Laboratories in USA clearly established that the matter which is exposed to sun's light should be a semi-conductor of the type silicon, selenium or germanium whose electrical properties are in between the electrical conductors (offering least resistance) and the insulators (that offer considerable resistance). Such materials discovered at that time were given the name 'semiconductors' and they are invariably non-metallic in nature.

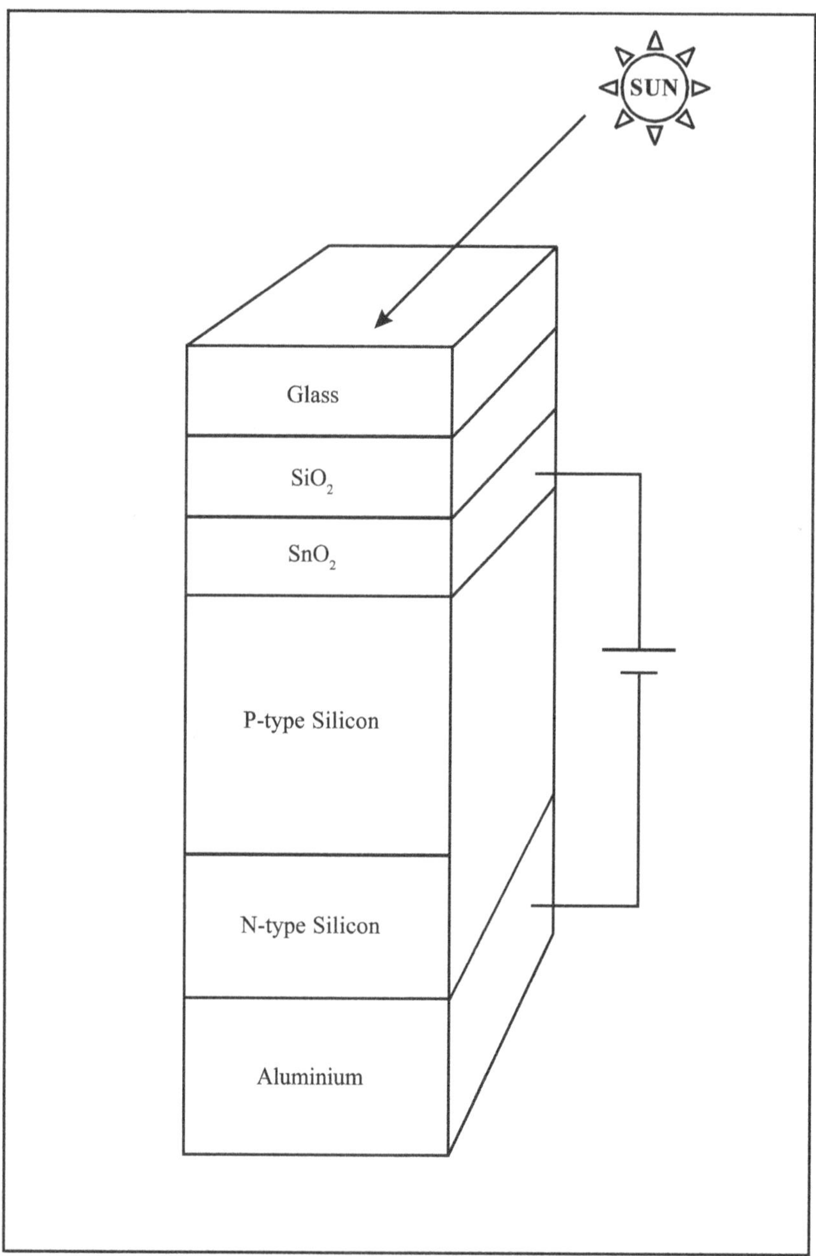

Fig. 5.8 *Solar PV Cell*

Thus by 1958 the first solar cells showing considerable efficiency, i.e., upto 16% conversion, have been manufactured on a large scale. The improvements in the fabrication of a cell took many years through the introduction of 'doping' or deliberate introduction of impurities such as boron, phosphorus etc., and slicing of the semiconductor into several thin films and construction of different forms.

PV Cells

A PV cell in its construction (see Fig. 5.8) is made up of two thin films of dissimilar semiconducting materials forming a clear junction. The semiconductor chosen for large variety of PV cells is silicon. Two types of silicon based semiconductors have been discovered; one, the n-type, made from crystalline silicon doped with phosphorous in small quantities so that the surplus of electrons from the *dopant* makes the whole semiconductor 'electron surplus' , with one of the surplus electron derived from phosphorus; the other, p- type, made from crystalline silicon doped with boron making it electron deficient in character. The missing electrons of boron thus cause 'holes' in the crystal structure and the character of having deficit of negatively charged electrons is finally termed as 'positive'charged silicon, p- type.

The 'original' crystal of silicon is cubic structured (valence 4) and each atom is shared by two electrons making a bond. The inter- atomic distance between each crystal is equal. The whole structure of silicon thus appear as a regular mat. Now a photon of light penetrates the PV cell near the junction, provide enough energy to excite one electron of a bond into higher energy level within the crystal. The electron thus promoted tends to migrate into the n-type silica layer and the hole into the p-type silica wafer. This process leaves current in the external circuit of the cell due to the movement of the electron, and later it enters the p-type wafer to recombine with the 'hole'. This is termed the 'photovoltaic effect'.

In the earlier experiments where high grade monocrystalline silicon was employed in the PV cells, very high efficiencies of solar energy conversions were achieved. These materials are made out of pure electronic grade crystalline silicon adapting sophisticated methods of preparation.

In recent times it became clear that such high grade silicon is not very much needed for routine uses of solar devices and a cheaper type e.g., polycrystalline or amorphous silicon can be used. Other simiconductors that find use are: gallium arsenide, copper indium diselinide and cadmium telluride.

Certain design techniques such as fabrication of multi-junction PV array, or light concentration on cells using mirrors to increase photon density will improve the functioning of the PV power generation technology. In the PV cells, the striking sunlight at the n-p junction produce a dc emf with p- terminal as positive and n- terminal as negative. The rating of a single PV cell exposed to sunlight are given as follows :

- voltage, 0.45 dc;
- current, 0.75 A dc;
- power, 0.33 W;

Solar cells are connected in series or parallel depending on the end use to give the desired voltage, current and power. When many cells are connected in series, a string is made; several strings if connected parallel form a module; and several modules when connected in series-parallel and so on form the configuration of an array or PV panel. The array is erected in slant to collect the solar radiation and convert it into dc instantaneously. Placing the PV cells array on roof tops in proper alignment also enhance the efficiency.

Utilities

PV cell systems find innumerable applications in day-to-day life despite the high cost and maintenance. Photovoltaic powered clocks, calculators and some weather equipment work on low power and are some of the utilities. Wide applications were envisaged in the development of power generation in remote mountain areas where power supply lines are difficult to be laid. Building lighting, street lighting and even village lighting are a variety of applications.

The Non-Conventional Energy Development Corporation of APLtd (NEDCAP) recently introduced into the civil market solar power lanterns, domestic lights, street lights, solar fencing, solar water pumps, solar cookers, solar driers and solar water heaters. ADITYA SOLAR SHOP, Secunderabad supplies and services all these items. The Indian Renewable Energy Develomment Agency Limited, (IREDA) under the Government of India, brought out several gadgets working on renewable energy systems and had been assisting upcoming entrepreneurs in establishing manufacture units by way of financial help. The NEDCAP of Andhra Pradesh in a proposal to Government of India emphasized the utility of 'solar blinkers' on road dividers, ghat section roads and near curvatures of roads. According to this, the PV cells that accumulate electric power during sunny times blink round the clock thereby giving caution to the drivers of vehicles and reduce the incidence of accidents. Also, we have today in the market several brands of solar water heating systems notably from PHOTON Energy Systems, Hyderabad, SUNRISE Solar Pvt.Ltd, Secunderabad, BHEL System marketed by UDAYARKA Renewable Energy Systems, Hyderabad, and a lot more.

PV driven automobiles were developed since 1980s by leading automobile manufacturers and the one demonstrated by General Electric Company is noteworthy.

The greatest use of PV systems is in the grid-connected town power supply. Kobern-Gondorf, Germany (340 kW), Phalk-Mont Soleil PV power plant, Switzerland (500 kW), The University of Northumbria, UK, Delphos, Italy (600 kW), and Sacramento Municipal Utility District (SMUD), California power plant (2 MW) are some of the gigantic ventures. A new concept of very ambitious grid-connected PV plant making use of the Satellite Solar Power Station (SSPS) concept by erecting 30 sq.kilometers PV cell arrays and generating electricity in space and converting into microwave energy to be transmitted onto the earth is expected to solve the power scarcity problems on the globe. However the cost of this project when conceptualised in the year 1972 was an unaffordable US$ 15000 millions; hence the project remained unfulfilled.

Questions

1. Define the solar constant.

2. What is solar insolation and how is this factor influenced by atmospheric and geographic features?

3. Which are the types of devices operated for solar energy utilization?

4. Give an account of various types of solar radiation collectors and explain how this energy is made to operate an electricity generation plant. What are the limitations in harnessing the solar energy?

5. Explain the three Rankine cycles that are theoretically applied towards generation of electricity from solar radiation.

6. Write the line diagrams for
   (i)   Solar thermal engine, and
   (ii) Solar heat pump.

7. Give an account of the photovoltaic conversion of solar energy. Which are the prominent semiconductors employed in this device and what are their characteristic properties to render electricity when exposed to solar radiation?

8. Write notes on PV cells. What are the applications in remote areas?

9. Describe briefly the methods of storage of solar energy (thermal) for subsequent conversions.

10. What is the principle of photovoltaic power conversion? Outline one power generation plant that is in operation in the world.

11. How is solar energy useful for the following applications?
    (i)   space heating
    (ii)   solar distillation
    (iii) rural street lighting
    (iv)   hydrogen production

6

HYDRO ENERGY

Hydro energy refers to the driving power that has come up with the motion of a vast sheet of water. This is derived by regulating the flow of water through construction of blocks known popularly since ages as 'dams'. The dams serve many purposes such as, storage of river water, diversion of a section of water, reservoir to serve the water for jobs to be performed later or for generation of electricity. Under the last category comes the use of dam for production of energy. The quantum of hydro energy produced in the world is approximately 5%.

Function

A mass of water moving either vertically or horizontally exerts kinetic energy. In many a case the water from the rivers flow through mountains situated at elevation and possess potential energy too. It is for this water to be held in a 'reservoir' a dam is constructed at an elevated spot. On the upper side of the dam, a water gate is operated to let water pass through a tunnel leading to the turbines at the bottom end.

Now, if the blade of a water turbine generator is placed in the path of water in motion down below, due to the force, the turbine moves on and spins the generators. From this it is imperative that the specific water head between the reservoir level and the turbine tail end determine the output generation of hydro power (Fig. 6.1). The three types of hydro power plants categorised on the basis of water head are :

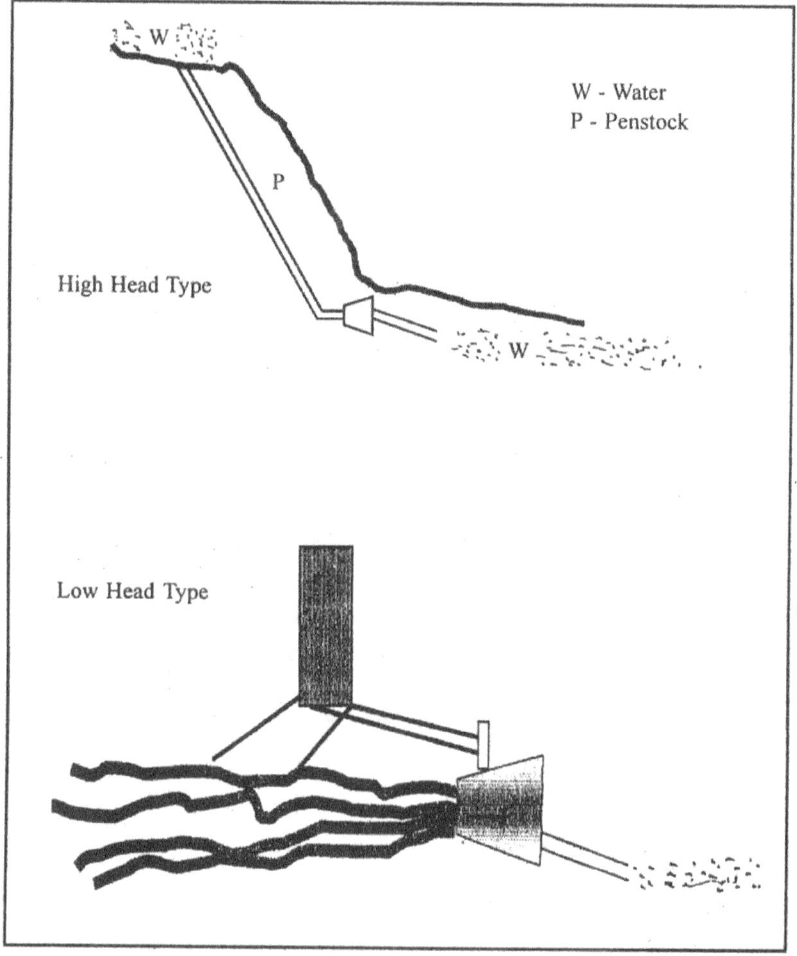

Fig. 6.1 *Categories of Water Head in Reservoirs*

High head	more than 200 m;

Medium head	30-100 m;

Low head	6-30 m.

Energy reserve in the reservoir is directly proportional to the height (h) of the water head and the quantity(mg) of water in the reservoir. Therefore a huge mass of water stored on a hilltop during the rainy season possesses a great energy potential.

The planners of energy systems(Central Electric Authority; Bureau of Indian Standards) deduced yet another classification of various hydro electric power stations based on the power production. Under this the following categories are listed:

Small hydro type	upto 6 MW;

Mini hydro type	upto 1 MW;

Micro hydro type	upto 100 KW.

Whatever be the classification, the outlook of the planners should be to develop all the hydro power stations at the natural ecological locations however small the water head is, irrespective of the quantum of power yielding capacity. Generally the range of these power sources are in the range of a few hundred kilowatts to several thousand megawatts.

Electricity can be generated from the power of flowing water. This can be achieved anywhere where the water runs down from a hill top or drop in from a dam or from irrigation canal backed by the water above-situated from a river. The hydro electric power plants generating electricity between 15 KW to 15 MW are the majority resources and are known popularly as mini hydro- or small hydro plants.

15 KW is the power requirement for 7 or 8 houses in an industrial country; it is also the requisite energy for a amall manufacturing plant or for about 50-80 small houses. About 15 MW power is the requirement for a medium-sized town anywhere and the plants producing this extent are known as 'conventional' or large hydroelectric plants. The line denoting SHP/MHP is not very clear and all SHP/MHP and Large Hydro Plants use similar machinery and work in the same way. From the data that is obtained, low head plants cost more as compared to SHPs. Thus one can attempt power generation for any water head ranging from 1 to 50 meters very successfully provided continuous assured water supply prevails. SHPs are thus the 'run-of-river' plants and produce cheap electricity that can be used in place of the thermal power. The water is backed up behind a dam or some diversion structure, where it enters the penstock leading to the turbine; here it forces the turbine to rotate and the turbine then turns the electric generator. The water, at the end, passes out through the draft tube into the tailrace and then back into the river/ canal. Electricity from the generator pass to the transformer, where the voltage is raised, and then goes through the circuit breaker to the power lines.

The topography of any hydroelectric station is as follows:

A dam / barrage built across the river is a barrier to control the flow of water. The reservoir or pool of water is generally on one side which is at elevation and the hydro-turbine generator is on the other side of the dam. It is seen that in all the three heads of the hydel stations the water head together with the quantity of water (also flow rate) determine the extent of power supply. The dam is usually constructed on a river valley and the water is held at upper level reservoir in the case of high head hydel stations. The water is allowed to flow through a pressure pipe to the low level turbine and after moving the turbine wheels it joins the tail end and finally the river. Medium head stations too have dams constructed on river but the turbine is situated at the lower level. In the case of low head hydel power stations the setting is s 'run-of-river' with a flat level high barrage across a river or even a huge canal with water flowing at high speed. A 'bulb type' or 'tube type' hydraulic turbine is axially placed within the passage of water.

A typical line diagram showing a SHP and a 'bulb type' hydraulic turbine is shown below for rapid understanding.

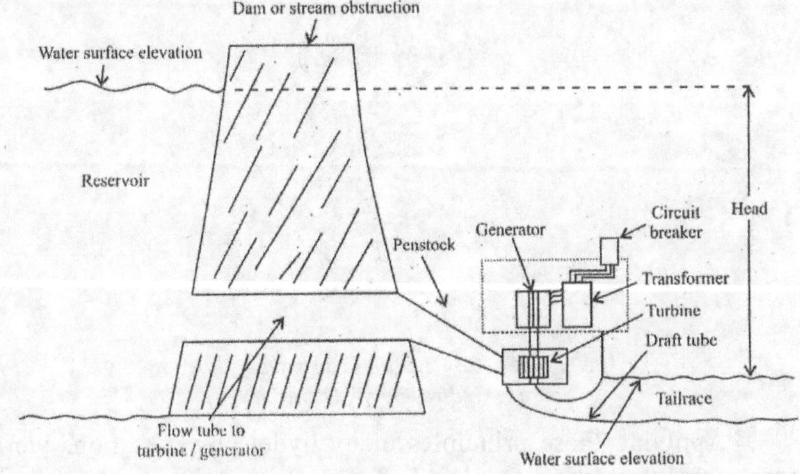

Fig. 6.2 *Typical mini-hydroelectric system showing major components and hydraulic head*

H = Head (in meters)

Three types of hydro turbines are in use at power generating stations. The Impulse type (Pelton wheel) receives water in high head systems. Usually the water is flowing down at high velocity and when this strikes the buckets of the Pelton wheel, the kinetic energy is converted to mechanical energy by the rotation of the wheel. Subsequently this mechanical energy is imparted to an armature assembly to produce electricity. The other two turbines are Reaction type (Francis and Kaplan) and in these, the water glides over the curved blades and pushes the blades. In all these processes, the kinetic energy of water is transformed into mechanical energy which in turn generates electricity.

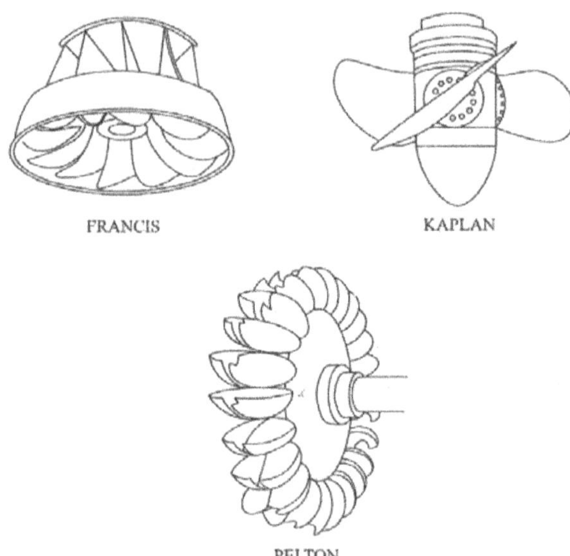

Fig. 6.3 *Hydraulic turbines*
Source : Godfrey Boyle, Oxford, 1996

Applying these principles many hydel power stations were constructed throughout the world. The electric power generated from the moving water of rivers is known as 'hydel power'. Every falling water system in our country was harnessed and electricity generated. This places the hydel energy resources in second position, the first being from fossil fuels. The first hydel power plant was erected in the year 1902 at Sivasamudram in Karnataka State and it was generating about 4500 kw power at that time. Subsequently, the Mettur, Bhakra Nangal, Hirakud, Tungabhadra, Damodar, Idukki and Koyna hydel power stations were constructed on several Indian rivers. The Narmada valley project and Tehri dam project are on the way. Still India has a vast hydroelectricity potential, this being 86,000 MW. Only 22,000 MW of this potential was targetted. The 10 major hydroelectric power plants operating in various Indian States and their power capabilities are shown in Table 6.1. The distribution of the power to various Indian states is being managed by the Central Electric Authority.

Table 6.1 Major Hydro-Electric Power Plants in India

Power Plant	Location	Capacity, MWe
Dehar	Rajasthan	990
Sharavathi	Karnataka	891
Koyna	Maharashtra	880
Kalinadi I	Karnataka	825
Nagarjunasagar	Andhra pradesh	815
Idduki	Kerala	780
Srisailam	Andhra pradesh	770
Bhakranangal	Punjab	710
Salel	Jammu & Kashmir	650
Kundah	Tamilnadu	555

Source: http://www.fe.doegov/international.indiover.html

The basic requirements for hydel power generation are: assured water supply for the river/ reservoir to sustain its potential energy; continuous flow of the water body in the mountain terrain at elevation; a vulnerable location to instal the turbine and receive the falling water; management of water that ultimately transferred the kinetic energy to the turbine e.g., by partial pumping of water back to reservoirs, and part supply into canals for irrigation. These requirements are mostly met in respect of Indian rivers which have a good potential for hydel power generation. India has so far set up 387 small hydro power plants(SHP) of upto 25 MW capacity to afford total power of 1341 MW. These are situated in hill regions as well as on canal drops. Even small water falls situated in mountain areas of Punjab, Hariyana and Uttar Pradesh were planned for power generation and the MNES database cites about 4100 potential sites that can yield a total 10,100 MW power. Even the participation of private sector in power generation is anticipated. It is expected that these SHPs are grid-connected for the purpose of better administration and distribution.

In Table 6.2, APGENO Power Generation details are given.

Table 6.2 *APGENCO Electric power genration*

Station	Generation (MU)		
	2003-04	2004-05	2005-06
Donkarayi Canal Power House (1 × 25 MW)	109.96	132	21.1
Lower Sileru Hydro Electirc Scheme (4 × 115 MW)	973.39	1171.3	186.2
Machkund Hydro Electric (Joint) Scheme (3 × 17 + 3 × 23) AP Share (70%)	828.2	900.6	147.8
Mini Hydro Schemes (Peddapalli 5 MW, Palair 2 MW and Chettipeta 1 MW)	12.3	7.6	0.3
Nagarjunasagar Left Canal Power House (2 × 30 MW)	0	5.1	0
Nagarjunasagar Right Canal Power House (3 × 30 MW)	0	47.7	0
Nizamsagar Hydro Electric Scheme (2 × 5 MW)	5	0	0
Nagarjunasagar Hydro Electric Scheme (1 × 110 + 7 × 100)	375.13	502.2	24.3
Penna Ahobilam Hydro Electric Scheme (2 × 10 MW)	0	0	0
Pochampad Hydro Electric Scheme (3 × 9 MW)	62.71	1.6	0
Singur Hydro Electric Scheme (2 × 7.5 MW)	7.7	1.4	0
Srisailam Left Bank Hydro Electric Scheme (6 × 150 MW)	327.6	1411.6	77.4
Srisailam Right Bank Hydro Electric Scheme (7 × 110 MW)	307.67	941	11.3
Tungabhadra Hydro Electric Scheme (4 × 9 MW + 4 × 9 MW) AP Share (80%)	114.6	148.3	0
Upper Sileru Hydro electric Scheme (4 × 60 MW)	399.83	544.1	79
Total :	3524.09	5814.5	547.4

HYDEL POWER STATIONS

While installing the power stations utilising the water flowing from a height to a basin at the down, one has to consider the working head (the column height) as it determines the potential energy of the water. The other parameters are the expected rated power output, the type of turbine and the actual location of the dam or reservoir. Even small volume of water stored at a high water head results in generation of considerable power as compared to large volume of water at a low working head. The water heads above 200 meters are regarded as 'high' heads, those between 30-100 meters are 'medium' heads and from 6-30 meters are 'low' heads. The water from high heads are brought down in a regulated manner through 'penstocks' which are conduits. The stream of water thus regulated and brought enters the plant and moves the turbine blades and drive the generator.

The water from medium and low heads is drawn in a comparatively easier manner through 'dams'. Once the infrastructure for installation of high head hydel power plants are made, the recurring cost of generation of electric power is very low.

MINI HYDEL PLANTS

Andhra Pradesh-Salient Aspects on Installed Capacity & Generation

Mini Hydro power stations are constructed to utilise water releases through irrigation canals.

Name of The Power Station	Name of the Canal	Name of the Resorvoir	Installed Capacity (MW)
Peddapalli	Kakatiya	Sriramsagar Project	7.6
Palair	Lal Bahadur	Nagarjunasagar Reservoir	2.0
Chettipeta	Godavari	Godavari Barrage	1.0
		Total	10.6

- Unit capacities vary from 220 KW to 1000 KW depending on the net head available for power generation.

- Eight mini hydro stations were constructed on D-83 Kakatiya canal near Peddapalli. First station was commissioned at 10th mile on 31.03.1986 and the latest addition was 7th mile on 01.01.2004. The remaining 18th Mile will be commissioned by end of January 2004

- Palair Mini Hydro Station (2 × 1000 KW) was commissioned on 13.02.1993 with a project cost of Rs.3.75 Crores.
- Chettipeta Mini Hydro Station (2 × 500 KW) was commissioned on 01.10.1991 with a project cost of Rs.0.88 Crores.

Generation Particulars

Year	Generation (MU)	Nagarjunasagar Reservoir Levels	
		Maximum (Ft)	Minimum (Ft)
2000-01	1522.71	565.4	508
2001-02	1067.22	576.1	497.8
2002-03	869.01	517.6	496.5
2003-04	375.13	--	--
2004-05	502.2	--	--

Nagarjunasagar Hydro Electric Scheme

Unit No	Capacity(MW)	Date of Commissioning	Make	
			Turbine	Generator
1	110	07-03-1978	BHEL	BHEL
2	100.8	08-04-1980	Hitachi-Japan	Melco-Japan
3	100.8	11-01-1981	Hitachi-Japan	Melco-Japan
4	100.8	22-06-1982	Hitachi-Japan	Melco-Japan
5	100.8	31-03-1983	Hitachi-Japan	Melco-Japan
6	100.8	26-10-1984	Hitachi-Japan	Melco-Japan
7	100.8	31-03-1985	Hitachi-Japan	Melco-Japan
8	100.8	24-12-1985	Hitachi-Japan	Melco-Japan
Total	815.6			

Salient Aspects on Installed Capacity and Generation

Nagarjunasagar multipurpose Dam is constructed across Krishna River near Nandikonda village in Nalgonda District. Main powerhouse is situated at the toe of the dam on left side.

Full Reservoir Level of Nagarjunasagar Reservoir is 590 feet and Minimum Draw Down Level for power generation is 510 feet. Gross storage capacity of Nagarjunasagar Reservoir is 408 TMCFt.

One 110 MW Conventional Unit was installed at a cost of Rs.15.60 Crores. Four 100 MW Units of Reversible Units were commissioned under Stage I of Pumped Storage Scheme and three similar Units were commissioned under Stage II.

Total project cost for commissioning seven Reversible Units under the above two stages was Rs.146.76 Crores. OECF, Japan has extended a loan of ¥ 15.4 billion for the above Scheme.

The special feature of the Reversible Units is that they can be operated in either Conventional Generator mode or Reversible Pump Turbine mode. They can also be operated in Synchronous Condenser mode to improve the Grid System voltage. Designed annual energy potential is 1230 MU.

Year	Generation (MU)	Nagarjunasagar Reservoir Levels	
		Maximum (Ft)	Minimum (Ft)
2000-01	1522.71	565.4	508
2001-02	1067.22	576.1	497.8
2002-03	869.01	517.6	496.5
2003-04	375.13	--	--
2004-05	502.2	--	--

Source: AP GENCO web site.

The construction of dams also regulate the water flow for irrigation purposes, control the floods besides generating the power. Some prominent dams in the world are: the Niagara dam, the Aswan dam, the Bhakra Nangal dam, the Hirakud dam, the Nagarjuna dam and the erstwhile Tehri dam.

Specialities of Hydel Power: All hydel power stations are constructed over rivers taking advantage of their course through the mountain/ hill areas. Facilities for pumping the water which has flown down into the basin to an elevated location into a reservoir are to be built up under the same project.

No air pollution is envisaged during the operation of hydel plants. Since no extra heat is generated in any systems, even thermal pollution is not observable. The hydel power is renewable and is only dependant on the rainfall that fill the rivers. The installations have a long life; even micro power plants of great durability can be erected within a short time of its conceptualisation. The cost of hydel power is very cheap. On one particular river many dams can be constructed for generation of power. The maintenance cost is also low. Excess water pumped into the reservoir utilising excess energy from the power plant can once again be made use of to move the turbines and generate electricity at will.

Limitations of Hydel Power Plants Installations: While siting the hydel power stations, vast agricultural and forest areas come under the operations. Moreover, disruption of hydrological cycle in and around the area as well as water logging occurs. The land salinity gets increased and siltation takes place in the reservoirs. Some storage capacity of the reservoirs also is diminished as silt occupies the tank. On the socio-economic front, the local population have to abandon the earlier professions and switch over to different ones much against their will. Some of these factors are playing a role in the execution of the Sardar sarovar project that is Narmada dam construction.

Questions

1. Give a brief account on the installation of hydroelectric power plants. What are the environmental impacts that happen during their installation?

2. What are the advantages and limitations in harnessing the hydropower plants?

3. What is the scope of building small hydroelectric power plants in India? What are the specific advantages of this type of plants over the high and medium hydel power plants?

7

OCEAN ENERGY

The ocean delivers three types of energy; thermal energy that is stored from Sun's radiation; kinetic energy derived from the motion of waves and tides. All the three categories of energy are practically clean. The infrastructural facilities required for erection of the plants are quite expensive and so the cost of power would be high; nevertheless the countries throughout the world have an open look at these ventures.

TIDAL POWER

The gravitational effects between the earth and moon, as well as the rotation of earth around the sun causes changes in the levels of seawater. As a result of these geographic phenomenon the seawater throughout the sea rises to great heights i.e., upto 20 meters in bays and open sea. The elevation of a mass of water to a height is termed as 'tide'. The formation of tides in seas is very systematic and this occurs twice a day with an interval of about 12 hours starting from the full moon day night with 'high tide'.

The water after remaining high in the sea for nearly six hours start moving on the surface layers in a slow fashion (known as low tide) and the change between the rising phase and slow phase of water is called the 'tidal change'. The term 'tidal range' (R) referred in the tidal energy denotes the difference between the consecutive high tide and low tide water levels; this value is denoted in meters. The amplitude of several ocean tides are between 25 cm to 10 meters and the speed of tidal currents in the range of 1.8-18 km/hour.

Tidal energy is converted into mechanical energy using hydro turbines. The generator's rotors are driven by this energy and thereafter electricity is produced in the same manner as a hydro electric system. Since the formation of tides are not continuous, it is possible to obtain power in an efficient manner only in intermittent intervals for short duration. In reality, tidal power is generated from a minimum water head of 5 meters and above only at some congenial locations in the world. A dam-like barrier for falling 'high tide' water, a water basin and a hydro-turbine are the components of the tidal power station. A line sketch showing the operation of a tidal power station is in Fig. 7.1.

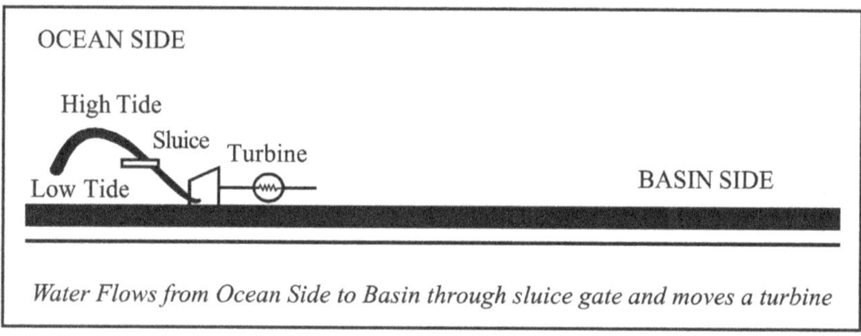

Fig. 7.1 Tidal Energy Convention

The technology of erection of these power stations is perfected nowadays. The cost of electricity generated from tides is very high on account of the development cost of barrage, land basin and the power generating machinary. Since tides formation in seas is eternal, tidal energy is renewable.

Tidal power availability is clearly site-specific throughout the world. The tidal power potential for the entire world is estimated as 120 GW (approximately 190 TWh/year). India's potential for extracting the tidal energy from its shoreline is rather low. So far no tidal power stations have been erected in the country owing to non-availability of technology; however the following power stations in other countries are generating electricity to the extent indicated against each of them.

La Rance river (France) = Tide av. 8.4 m;
349 MW; erected in 1966.

Fundy Bay (Canada) = Tide range 5.5-10.7 m;
722-10,000 MW; erected in 1984.

Lumbovskii bay (Russia) = Tide av.5 m; 277 MW.

However the locations of the sea at which tidal power stations can possibly be built in India are:

Bhavanagar (Gulf of Cambay)-Average tide 6.8 m and expected power generation 5510 MW.

Kandla (Kutch)—Average tide 5.3 m and expected power generation 1182 MW.

WAVE ENERGY

Wave energy is purely a kinetic energy which the river or ocean develops under natural conditions. The waves over water bodies are created by the pressure of the wind thrust on them. A wave is a transition of kinetic energy; that means, the wave action starts at a far reaching place and on account of the motion, the energy of the wave can be felt at a different location which is quite apart. It is not all the 1.5% of solar energy that got converted into wind energy generate the waves, but only a small fraction of the wind energy that contribute towards its formation. Waves are perennial in seas and large rivers. Waves possess both potential and kinetic energy. Wave height which is calculated as $2 \times$ amplitude is generally in the range of 0.2-4.0 m and wave period in the range of 4-12 seconds.

The energy can be tapped from the moving waves by operating a wave machine. The wave machine comprises of an air turbine which is floating on the surface of water receiving the disturbances of wave motion and rotating in itself. A small concrete caisson which is laid in the path of the waves acts as a break-water. Inside the caisson are a small column of water and entrappod air. Now, the energy of waves is transferred to the small column of water; the pressure of air increases and launches motion in the air turbine thereby operating the generator. Electricity is generated in the machine.

The principle involved in the generation of electricity from waves is illustrated as follows:

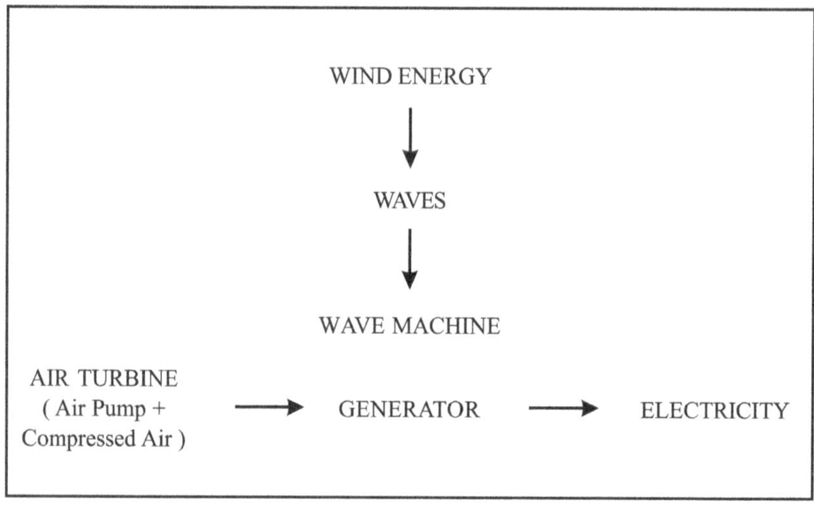

It could be infered from the above that the atmospheric wind hits the water in the vast sheet of river, sea or ocean. It transfers its potential energy to water which generate the so-called ' waves'. These waves have the energy within them and it is possible to extract some of its energy by striking an 'oscillating water column' (OWC). In this process the air inside the OWC is compressed and made to move an air turbine to generate electricity. A line diagram of Well's air turbine which transforms wave energy into direct current is shown in Fig. 7.2.

The first application of wave energy was made in Japan to power a lamp on a buoy laid in sea. Later the Norway wave device installed at Tofleschallen was producing 500 KW electricity in operation till 1998 when it was blown off in a storm. The Islay wave power generator was subsequently constructed with the efforts put by WAVEGEN and Queen's university, Belfast. Wavegen developed three types of energy generating devices. these are:

- LIMPET 500, a 0.5MW shoreline power station designed to operate on exposed shores for local or island or port power supply.

- OSPREY 2000, a 2 MW near-shore gravity anchored power station to cater the region.

- WOSP 3500, a 3.5 MW near-shore combined wave and wind power station.

Basic Design

The two main components of the devices are:

1. ***The Collector and Oscillating Column:*** This is a shell-like structure into which seawater freely enters or leaves causing different levels of water in the shell. Now, a column of water with air over, that is in continuity to the shell , is alternately compressed or decompressed by the movement of seawater. This situation creates an alternating stream of high velocity air in an exit blow-hole. This air under pressure causes motion in an air turbine thereby generating electricity.

2. ***The Turbo-Generator:*** WELLS air turbines have the unique property of turning in the same direction regardless of which way the air is flowing across the turbine blades Fig. 7.2a. The turbines operate on both the rising and falling wave levels that take place in the collector and continuously drive the generator to produce electricity. A line sketch of the wave energy generator is shown in Fig. 7.2b.

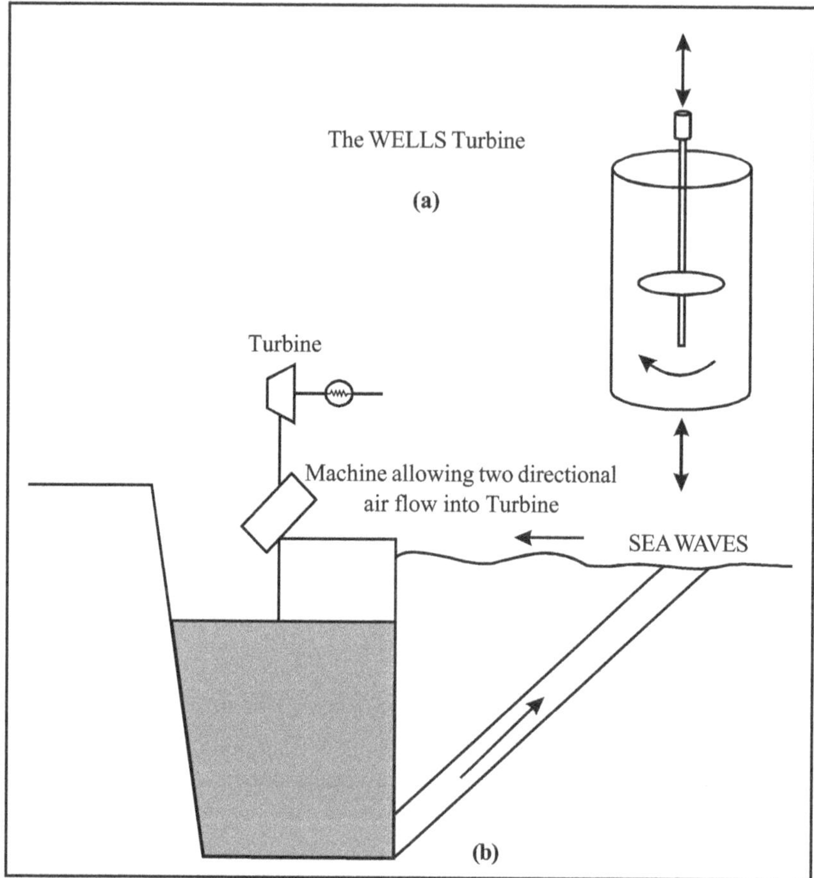

Fig. 7.2 (a) The Wells Turbine (b) Wave Energy Conversion

An experimental wave energy conversion plant expected to generate 150 KW was commissioned near Trivandrum, India. It was a multi-resonant OWC installed in a breakwater employing a 2.0 m drain Wells turbine and induction generator. Since the cost of the breakwater is shared between the Harbour and Power Plant sectors, the cost of power generation may be economically feasible. However in reality, the unit is delivering an average 75 kW output from April to November and 25 kW from December to March of the year. The R & D efforts to construct a few more OWCs around Indian coasts are in progress.

OCEAN THERMAL ENERCY CONVERSION (OTEC)

Oceans cover more than 70% of the earth's surface; The surface waters of the oceans absorb large quantity of heat from the solar radiation and store it. The surface waters are at higher temperatures than the deep waters and a temperature difference of 20 $^\circ$C persists at some locations near coastline. It is thus possible to extract this heat energy from the surface waters and convert the same into electricity. There is a great potential to extract this energy throughout the world. It is estimated that the world's share is to the tune of 10 million MW and India's is about 20,000 MW

There are two types of systems operating in the world. These are:

1. ***Open Cycle (also known as Claude cycle, Steam cycle):*** It employs the seawater to generate the steam required for moving the turbine. In an evaporator which is subjected to vacuum, surface seawater entering this space immediately gets converted into steam. This steam is in turn made to drive the turbine and produce electricity. Cold water was piped up from lower ocean depths to cool the steam (into water) that has come out of the turbine. In this way the cycle begins again. Adapting this design George Claude in 1929 constructed the first OTEC device and produced 22 KW electricity off the coast of Cuba .

2. ***Closed Cycle (also known as Anderson cycle, Vapour cycle):*** In this system a working fluid of the type, ammonia or butane or freons, is used to extract the surface heat of ocean water and vapourise and drive the turbine to generate electric power. Closed cycle OTEC is very similar in function to a 'heat engine'. Line diagrams of open and closed cycle OTEC are drawn in Figure 7.3.

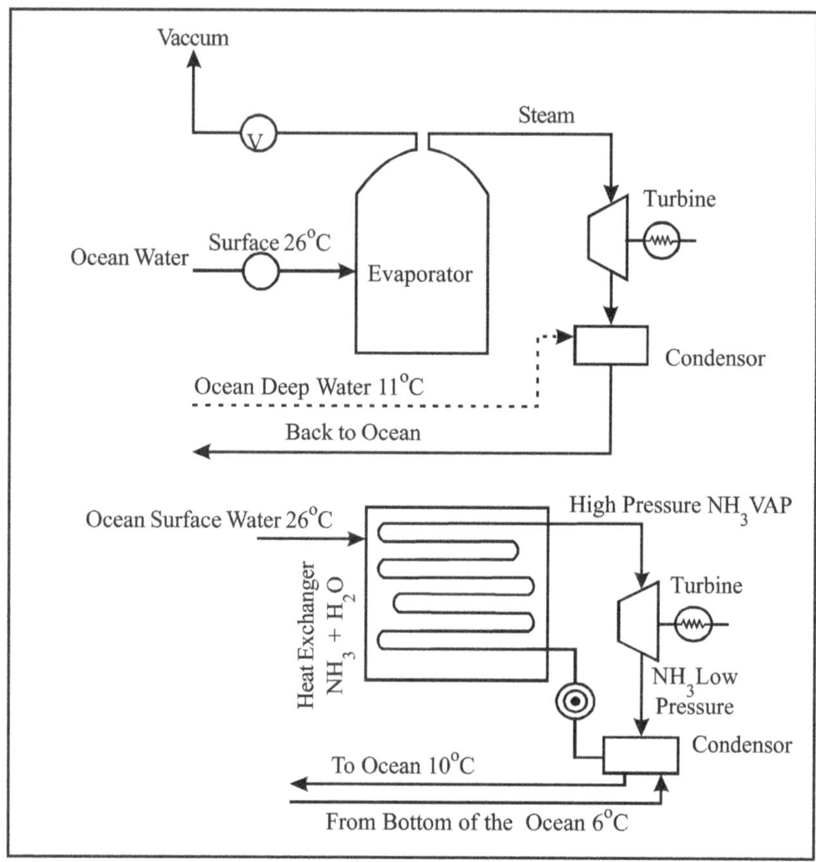

Fig. 7.3 OTEC Open & Closed Cycle Plants

A specific advantage of open OTEC design is that the generated steam in OTEC is at very low pressure and high voume as compared to the conventional steam power plant. The working of such an OTEC plant should be relatively simpler.

Several designs for construction of Ocean Thermal Energy Converters were drawn but they are still in the development stage. Even the cost of erection of OTEC is high. Japan and Hawaii coastal sites were chosen for the siting of several converters to give electricity in the range of 50-100 KW. A proposed OTEC to produce 1 MW electricity is recently conceptualised and is to be executed shortly in Tamil Nadu.

Questions

1. What is tidal power? How is tidal power harnessed at the basin of sea to generate electricity?

2. What are the essential requirements for establishing tidal power stations? Which stations in India fulfill these specifications?

3. How are waves generated in the water bodies? Explain the phenomena of producing electricity from the waves using 'wave machine'.

4. How is the heat from ocean extracted and utilized to move the turbine for producing electricity?

5. Explain the Claude cycle and Anderson cycle methods for generating electric power from the heat of ocean.

8

WIND ENERGY

The difference in incidence of solar radiation on earth surfaces causes variation in heat and thereby the heating-up of surrounding air. The air in the hot region of the earth expands faster and builds up high pressure and similarly the air at the less- hot region produce low pressure. This situation of pressure gradient causes movement of air from high pressure region to the low pressure area and this phenomenon is known as 'wind'. The formation of winds are continuous as the hot air on earth tends to move upwards allowing cold air to reach the surface of the earth in a cyclic manner. The rotational force of the earth determines the 'wind direction'. Wind possesses kinetic energy and this is to an extent of about 2% of the solar energy incident on the earth.

Wind power was harnessed since early times for useful purposes such as driving and directing small vessels over rivers, pumping the water from wells, grinding the paddy in fields and cutting the wood through levered / geared saws in rural areas in India, Egypt and European countries.

However the industrial revolution in Europe facilitated the discovery of generating electricity from the wind power since the year 1925. Thousands of devices, known as 'wind mills' came into operation in Germany, Switzerland, Denmark, England and Holland since then.

A distinction between 'wind power' and 'wind energy' has to be made at this stage. Wind power is actually the ability of the available wind to do work (which may be much more) and wind energy is the measure of the actual work done (which is partial) when any wind-powered device is operated. Wind power is thus expressed in terms of electrical power, Watts per square meters of air drift. Thus the available wind power at many locations in western countries has been established and some important conclusions drawn. A location where wind power estimate shows lower than 100 Watts per sq.meter is unsuitable. Areas with available power of 100 to 200 Watts per sq.meter fall under moderate zones and those exceeding 200 Watts per sq.meters are considered the best locations.

Again, these parameters are to be integrated to the wind speeds at the region and final assessment made. It has been experimentally established that the wind power generated is proportional to the air density, the area that is swept out by the wind device (ie. the blade length), and the cube of the speed of the prevailing wind The natural winds moving at higher speeds therefore produces greater energy as compared to winds of lower speeds, the result being eight fold. One empirical generalisation drawn from these studies is that a minimum wind speed of 16 km per hour is required for establishing an effective wind mill at any location. The kinetic energy of the wind is partly converted into the mechanical energy of the rotor (to an extent of about 60%) which in turn transforms into electric power (to an overall 30-40%).

WIND ROSE

A 'Wind Rose ' is a pictorial design on board that depicts the information on wind direction, speed and prevalence of calm conditions over a specified zone during a particular season or the whole year (Fig. 8.1).

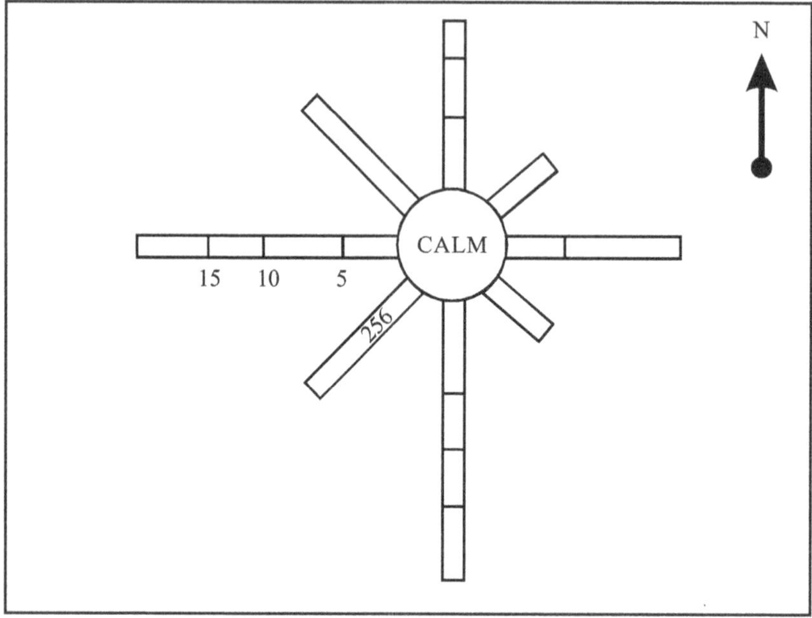

Fig. 8.1 *Wind Rose*

The basic data needed for drawing the wind rose is obtainable from the local meteorological stations. Once the wind rose diagram is drawn, it is possible to interpret the atmospheric pollution and weather data, or selecting sites to install wind mills or any other type of activity.

Method

The diurnal wind speed, direction data at a selected site 'X' is collected for a period of at least five years. An eight directional cycle wheel is drawn and the directions, namely N,S,E,W, SW,NW,NE and SE are marked as the arms. The different values of wind speed are segregated under the categories 0-5 mph, 5-10, 10-15, 15-20, 20-25 and so on for the entire five year period. Now, each of the arms showing the direction is scaled in a linear fashion as 0,5, 10, 15, 20 and 25 and a box is created against each of the category. At the centre of the wheel a circular box is drawn.

The figures written within the boxes indicate the number of occasions on which a wind of that particular direction has attained that specified speed. All occasions when the wind speed values shown were less than 5, irrespective of the direction , were pooled together and placed in the central circle, which depicts 'calm' condition. A wind rose drawn in this fashion offers at a glance, the nature of the wind in the area for years to come and will be a guideline for establishing a wind mill.

WIND MILL

The system which harnesses the speedy wind into electricity is known as 'wind mill'. Wind mills work very efficiently at uninterrupted locations, generally at the tip of mountains or on high towers. Some of the general guidelines for siting the wind mills are discussed here. The first step is to gather meteorological data pertaining to wind speed and direction at the location of interest. This is usually obtained from the nearest weather station. Seasonal 'wind roses' are then drawn for ready reference. This will help in deciding whether the requisite wind potential exists at the location. If the wind pattern, i.e.,wind speed and direction are apt for construction of wind mill at the site, further preparations are made. Generally the tower of the mill is raised at a location which is at least 30 feet higher than the nearest obstacle situated in a 300 feet radial zone. Places where tall trees or buildings are situated in the neighborhood are avoided. Hill tops where the turbulences are minimum are more convenient than the slopes. A reasonable height of 60-80 feet has to be the tower elevation. Topographical elements such as hills, trees, valleys and water sources which are in the direction of the wind influence the speed. The properties of sea winds and mountain winds are also to be taken into account while siting a wind mill.

Wind Mill Components

The three structural components of a wind mill are:

- The Tower,
- The Rotor,
- The Generator.

A description of various types of these systems and their functions has been made in the following paragraphs.

The Tower

The tower gives a good show of the wind mill. It is grouted on the earth firmly, generally over an area of approximately one fifth of the tower height. A 80 feet high tower should normally be grouted in a floor area of 16 square feet in the earth. Steel is the material of construction of the tower. Soft metals should not be used in the fabrication of towers. Towers have to be robust to withstand the speedy winds, bear the load of rotor and ancillary machinary. It should be vibration-proof and withstand shocks. It should regulate the turbulence of wind also. Never a tower should be positioned on the roof tops of buildings. Towers have to be painted with corrosion resistant paints and maintained in good condition. In Figure 8.2 are shown three types of towers which are in present day use. These are

1.   Pole tower,
2.   Lattice tower
3.   Guyed tower.

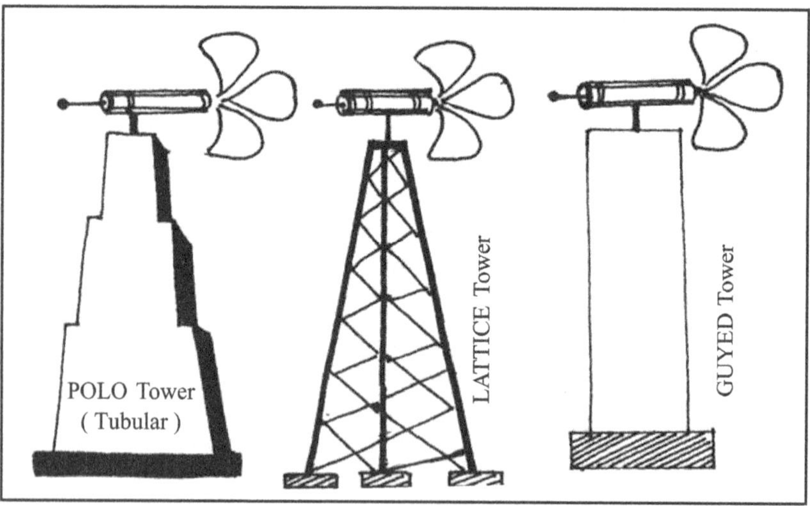

Fig. 8.2 Wind Mill Towers

The Rotor

The rotor is an important system in the wind mill. It actually determines the quantum of prospective energy that comes out from the mill. It must turn the generator at an optimum speed to generate maximum power. A rotor consists of blades arranged in a radial manner. The blades are set at a slight angle in the wind direction to facilitate fast rotation. Usually the outer tip of the blade attains five times greater speed than the actual wind, and fans lot of air for causing rotation. It means that if the wind speed is 12 mph the rotational tip speed of blade is 60 mph. The rotor tip speed ratio is thus–the ratio of tip blade speed to actual wind speed. It is observed that a rotor tip speed ratio of 5-8 affords uniform and reasonable power generation. Furthur, the rpm of the rotor is dependent on the length of the blade, the longer the blade the lesser the rpm it makes. The power output from a large blade rotor is considerably great as compared to small blade-constituted rotor. Power output is proportional to the square of the length of the blade. Rotor blades are made up of stainless steel, FRP, steel, wood, aluminium and other light alloys, the criteria being light and sturdy.

The arrangement of blades of a rotor in space is generally in two forms,

1. Horizontal axis and
2. Vertical axis.

When the axle (shaft) of the rotor carrying the blades is in parallel position to the horizon, the system is called 'horizontal axis' type. If the axle is in vertical position to the horizon, it is known as 'vertical axis' type. These forms of rotor positions are shown in Fig. 8.3.

Most of the wind mill rotors are designed to work at wind speeds between 8-40 mph. Sometimes overspeeding of rotor takes place and this will spoil the turbine. Also, it is not possible to predict the impacts of weather at short notice. Therefore adequate automatic controls are incorporated in the rotor systems to orient in such a manner as to avoid the greater wind speeds. These are achieved following the techniques such as:

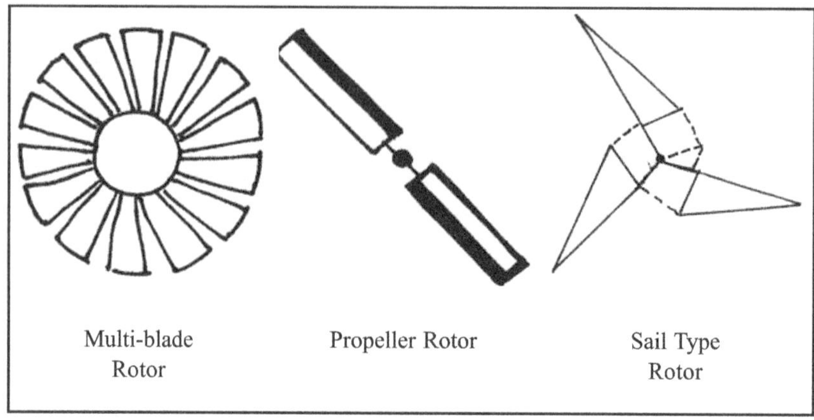

Fig. 8.3 *Wind Mill Rotors*

- Variable axis control- turning the rotor sideways or upwards away from the wind.

- Coning speed control- wherein the blades are actually made to bend in cone form due to wind.

- Spoiler flaps- by ejecting flap that slows down and acts as a brake.

- Blade pitch control- by the action of weights that change the pitch of the rotor.

- Brakes- by mechanical means.

Both horizontal axis type and vertical axis type rotors have reasonable aerodynamic stability while functioning with the adequate fittings.

The Generator

Wind power can be used to generate either DC or AC electricity through the operation of suitable generators. The principle of operation of a DC generator is as follows: A coil of wire rotates between the north and south poles of a magnet. The coil gets inducted with electric current and it is regulated at the end of the coil in a commutator segment to flow in a single direction. The design of an AC generator is similar to the DC generator except that the commutator segment at the end of coil is changed by copper rings, called 'slip rings'.

In practice, an ac generator has the armature (coil, i.e., stator) fixed and the field poles (rotor) rotated. The ac current is generated when the coil is kept in perpendicular position to the field of the magnet and as rotation occurs, current changes direction with each cycle. In Fig. 8.4 are shown the components of a dc generator.

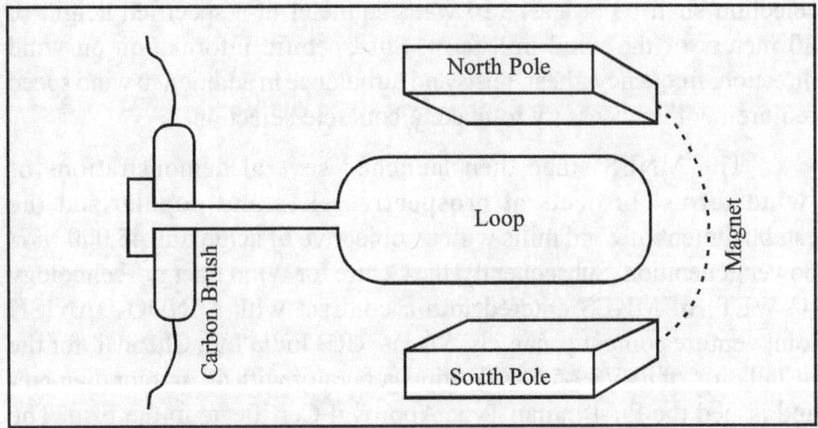

Fig. 8.4 Line Drawing of a DC generator (moved by wind)

WIND FARMS

The construction of a series of wind mills in an elevated zone and connecting the power generated from the entire area constitutes a 'wind farm'. While exploiting this non-conventional energy, this group planning has to be made for economic reasons. In the state of California, USA, wind energy is produced through an agglomeration of wind mills of the farms. The wind farm at Lamba in Gujarat state has an installed capacity of 10 MW.

FEATURES OF WIND ENERGY

Wind is a renewable resource, available free of charge. It is plenty throughout the world. During the production of this energy no pollution of air or water occurs. The technology is well developed and proven. Necessary machinary for harnessing this power is available commercially.

As regards India, the wind data assessment of several sites in states was made by the Indian Institute of Tropical Meteorology, Bangalore, in association with the Ministry of Non-conventional Energy Sources(MNES) and the list of suitable sites where the average wind speed throughout the year exceeded 18 kmph was published. Furthermore, it was concluded that the annual mean wind power density at the site of selection shall be at least 150 watts/sq.meter at a specified height of 30 meters of the wind mill/farm. Site-specific information on wind direction, frequency, shear, gusts and turbulence in addition to wind speed requirement is necessary to make a concrete selection.

The MNES since then launched several demonstrations of 'wind farms' projects at prospective sites and popularized the establishment of wind mills with an objective of achieving 45,000 MW power generation. Subsequently, the Centre for Wind Energy Technology (C-WET) of MNES entered into a contract with a INDO-DANISH joint venture company, namely, Vestas RRB India Ltd, Chennai, for the installation of its V39-500 KW wind generator with 47 m rotor diameter and issued the First Indian Type Approval Certificate to the firm. The Vestas Type WEG, a popular design throughout the world, has thus entered the Indian soil and to day, the company installed several Vestas WEG Type wind mills is the states of Gujarat, Maharashtra, Madhya pradesh, Orissa, Tamil Nadu (first installation), Kerala and Karnataka. In India, Vestas RRB company offers two types of wind electric generators, namely, Vestas Type V 39-500Kw with 47 meter rotor and VestasType V 27-225KW. The former is considered to be more suitable to the Indian wind conditions and seasons. In conclusion, it may be stated that the technology and logistics for harnessing wind energy in India is fully understood and applied.

Name & Address	Capacity (MW)	Date of Instalation
Ramgiri Wind Farm, HLC Colony Anantapur-515001 Ananthapur(Dt.) Phone : 08554-276541 Fax:	2	10/10/1994
Total Capacity:	2	

Wind farms can be located where strong dependable winds are available most of the year. On account of the uncertainity of strong winds throughout, generation of wind power is intermittent. Therefore this type of energy has to be either augmented with the regular town power supply or used as a stored energy, e.g., chemical energy in batteries, or generation of high energy material- hydrogen.

The noise produced from the mills is annoying and sometimes interfere with the radio waves. Even the aviary is affected in these regions. Lastly, wind energy is costly as the infrastructural framework is expensive.

Questions

1. How are winds formed in the atmosphere?

2. How is wind power harnessed to produce electricity?

3. Explain the components of the wind mill and how does it operate to move the turbines?

4. Write the salient features of 'wind rose'.

5. Discuss the potential of installing wind power stations in Indian states.

9

GEOTHERMAL ENERGY

The planet earth was believed to have been formed about 5 billion years ago upon a collision between vast clouds of dust and gases in the space. It was conceived that the planet assumed tremendous heat within it , especially in the core. The heat at the outer surface appeared to have cooled down over a period of time, but the inside remained still hot. The Geothermal Energy which we receive from the earth is the natural heat stored beneath its surface. Inside the earth the temperature rises with depth with a gradient of about 30 C/km. A lay-out of the earth's layers are shown in Chapter 1, Fig.2.

The inner core of the earth is the hottest region which is estimated to maintain a temperature of 6000°C. It is a red-hot solid body made up of metals, and alloys of iron. Over this is layered the outer core, which again is constituted of molten metals. Some of the radioactive metals such as thorium, uranium, isotopes of calcium, strontium and potassium etc.,are also present in these molten fluids.

Now, a solid rock mass which is having fairly low thermal conductivity, called mantle is surrounding the outer core. Over this exists the *crust* that takes practically no heat from within, except when eruptions of volcanoes or crevices occur. Human settlements and the objects of environment are set on this crust of the earth.

In some places of the earth, the crust is rather very thin. The molten rocks down below this layer (mantle) erupt out in these areas and spread as volcanoes. The hot magma of molten rock is usually at a temperature of 1260°C and this is one form of geothermal energy. In some cases water is trapped underground in the hot rocks. The water exchanges the heat from the hot rocks and becomes steam. The steam along with hot water springs up to the surface of the earth resulting in the formation of 'geysers'.

Geothermal heat is a natural form of energy the man knows throughout the world. In the Canary Islands, if one scraps the soil with the hand, the skin would be burnt. The geysers in the Yellowstone National Park shoots up hot water at intervals of an hour to a height about 200 feet. At Svartsengi, Iceland, local people swim in the 'warm lake' of the geothermal station eventhough outside is pretty cool. English Tom Bath derived its name from the experience of taking bath in the Roman-built bathhouses getting hot water from geothermal sources. Ancient Polynesian Settlers in New Zealand were living on the geothermal resóurces of water for cooking, bathing and washing purposes for as long as 1000 years without disturbance from the European monopolists.

In the earlier days the energy has been directly utilised for warming rooms, water in ponds or swimming pools or bathing. Even today, the hot springs or geysers have been utilised as health resorts in India, Italy, Iceland, Japan and some European countries. Power plants based on geothermal energy were set up at Rosemanowes, SW, England, Yellowstone geysers, and San Fransisco, USA, Wairakei, New Zealand, Larderello, Italy, Big Island,Hawaii and Namafjall, Iceland. The total energy produced account for 0.8% of the world's energy generation. Currently over 8000 MW electricity is generated in 18 countries of the world. About 11,300 MW equivalent of geothermal heat is directly utilised in the world.

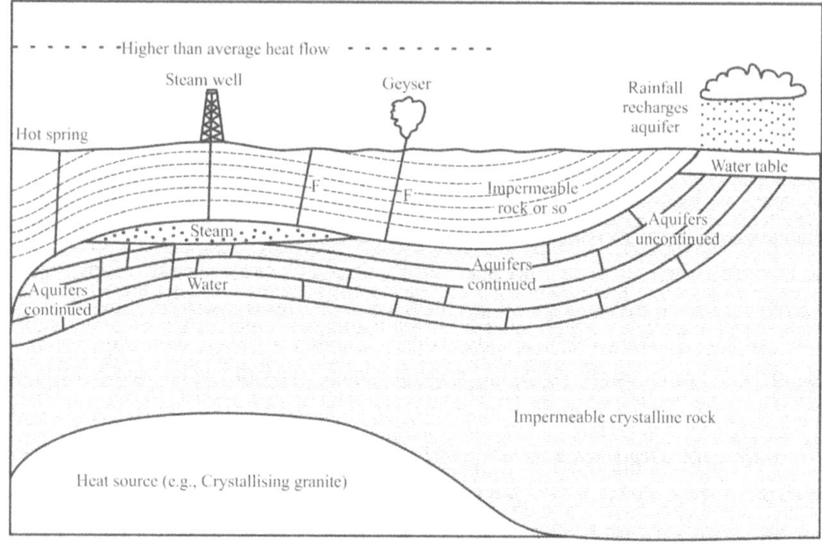

Fig. 9.1 *Requisites for generation of G.T.E*

Source : Godfrey Boyle, Renewable Energy, Oxford University Press, Oxford, 1996

There are different scenario through which the geothermal energy is tapped. First, the geothermal setting of the region; second, the hydrological parameters at the location. Geothermal resources in most types must have three important characteristics. These are: an aquifer containing water that can be accessed by drilling; a cap rock to retain the hot fluids and a hot rock. Aquifers are porous rocks e.g., porous limestone through which water freely flows and stored at times. Over the aquifers, mudrocks, clays and unfractured lavas that settle down over ages constitute the cap rock. The most important requisite, the hot rock, is either a natural magmatic intrusion set by an extinct volcano in the form of crystallising rock (partially molten rock) or hot dry rocks lacking in aquifer. Steam or hot water often ooze out through the fissures or faults of the cap rock. Wherever steam alone is oozing out to the surface of earth, the exit is known as 'fumarole'; geysers are those where steam and water are collected; hot springs are those giving out only hot water (see Fig. 9.1).

EXTRACTION OF GTE

Hydrothermal System: This is a reservoir of dry steam. The hot source down the earth is covered by a permeable mass through which water enters and comes in contact with the hot body. Instantaneously, dry superheated steam is produced and due to the high pressure in the enclosed space it finds way to the surface of earth under high pressure (usually 30 atmospheres). Some of the rock mass and impurities in the steam are eliminated and subsequently it is driven into the turbine to generate electric power.

Geopressure System

The ground water in the earth comes in contact with hot rock, magma. and gets heated to temperatures upto 700°F(180-350°C). This water-steam is drawn to the surface through digging a well from the cap rock. At the ground, when the pressure is released, a part of water turns into steam and that is led into the turbine for generating electricity.

Hot Water System

At many locations on the surface of earth, hot water at temperatures ranging from 50-80°C spring up continuously. The heat preserved in this water can be directly utilised for room warming or after exchanging with low boiling organic solvents, it can be converted to electric power.

Hot Dry Rock Systems

There are many areas in the world where the formation of hot rock close to the surface of earth prevail, but significant quantities of ground water in the vicinity is absent. Therefore the heat of the rocks which is usually at a temperature of about 290 °C cannot be immediately extracted. In such cases, two external deep holes are drilled at a distance of

100 yards apart in the rock bed. Through one hole (injection well) large quantity of water is injected to fill up the interior rock volume thereby extracting its heat. By means of a pump the hot water that is created in the rock bed is drawn out to the surface of earth through the other hole (extraction well) and utilised to generate power. The condensed water is recycled in the operations of the two wells.

Based on these systems, technologists have conceptualized two cycles to harness the geothermal energy for power plants. These are :

1.   Flash cycle, which extracts the heat of earth at temperatures above 150°C and

2.   Binary cycle, which operates at temperatures between 90 -150°C.

The flow diagrams of these two cycles are drawn in Fig. 9.2.

Flash Cycle

In this cycle the hot water (also termed brine) which is obtained from the geothermal heat exchange is flashed in a tank, whereupon the pressure of the fluid is reduced rapidly resulting in the formation of steam. The steam is then led into the turbine to produce electricity.

The condensed water from the steam as well as the unvapourised geothermal fluid are reinjected into the well. This is a continuous cycle starting from the injection of water into the earth.

Binary Cycle

The hot geothermal fluid exchanges the heat with a low-boiling, stable working fluid in a heat exchanger. The working fluid in turn goes through the cycle to move the turbine and generate electric power. Fluoro hydrocarbons of the freon types are generally employed as the working fluids.

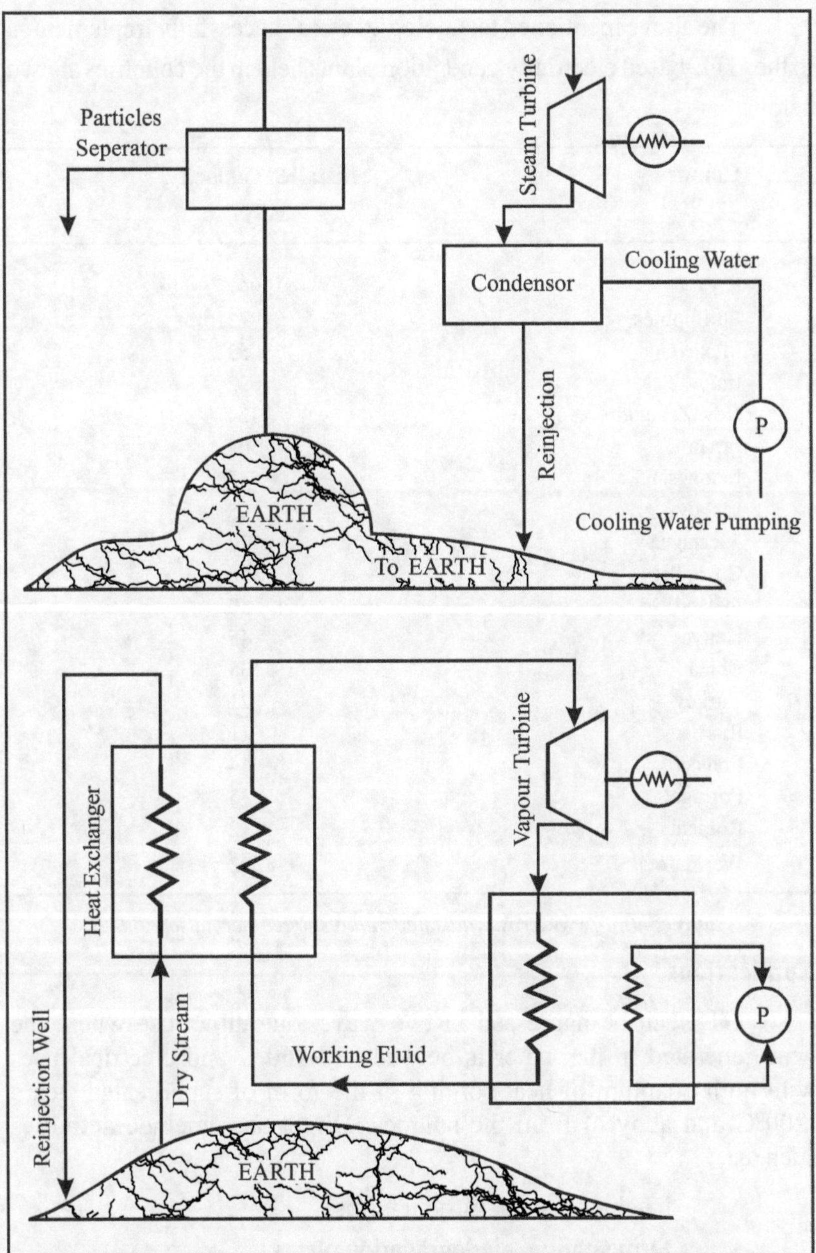

Fig. 9.2 *Line Drawing of GT Power Plants (Flash Cycle & Binary Cycle)*

The above mentioned technologies were successfully implemented in the GTE-based electricity generation plants held in the countries shown below:

Country	Installed Capacity, MW
USA	2961
Philippines	1922
Mexico	753
Italy	632
New Zealand	324
Japan	504
Indonesia	587
El Salvador	105
Nicaragua	70
Costa Rica	60
Iceland	50
Kenya	45
China	35
Turkey	21
Russia	11
France	4
Portugal	5
Romania	2
World Total	8392

Source: http://www.crosswinds.net/^nqureshi/geothermalscan.html

Applications

GTE can be harnessed in two ways; one direct use where the heat generated in the water is between 30-160°C and electrical uses with high quantum of heat coming in the form of superheated steam (200°C and above) from the sources. Direct uses include activities such as:

- hatching fish, poultry in farms,
- fermentation, biodegradation plants,
- swimming pools,

- refrigeration and space heating devices,
- drying chambers for organic materials, seaweed, grass etc.
- evaporation processes e.g., sugar solutions,
- drying of diatomaceous earths and minerals.

Heat Pump

The general design of a heat pump is described under solar energy, Chapter 5. GT Heat pump is similar to that, and it consists of a well-laid pipeline buried under shallow depths of the earth (>20-30 feet) nearby to the building, a heat exchanger and ducting into the building. As well-known, during the winter time, the interior earth is slightly at more temperature than the outside. This heat from the relatively warmer earth is drawn by the heat exchanger and delivered to the indoor environment in the house. In a similar way, during summer, the heat from the hot air of the building is exchanged by the heat exchanger and is pushed into the relatively cool earth. The heat extracted from the building during summer is ultimately used up to warm additional water, thus maintaining an equilibrium of cool temperature between the building and the earth. Heat pumps of varying sizes and capacities making use of geothermal heat for small and large buildings have been designed .

Direct Use Designs

Many systems designed for direct use of GTE are based on three components. These are:

- A geothermal well, to deliver hot water of required temperature or heat.
- Conduits and heat exchangers to absorb the heat and pass on to the space facility.
- A collection system of water returning from the space facility, i.e., an injection well or a storage pond to receive the returning water/ fluids.

Suitable designs are adapted for these utilities.

The Geological Survey of India in 1973 identified 340 potential sites for tapping the geothermal energy. Many of them were expected to render fluids at temperatures 100-120 °C but a few wells at depths of

1-3 km were expected to produce fluids at temperatures of 200-250°C. A 5 KW binary cycle plant was successfully installed at Manikaran field of Parbati Valley in Himachal pradesh, in 1992. Two geothermal power plants, one at Puga-Chumathang (Ladakh district, J&K, 1MW capacity binary cycle plant) with 115 springs, total discharge 18 kl/h, temperature 50-84°C and another at Tattapani region in Madhya Pradesh (20 MW capacity binary plant) with 23 springs, total discharge 3600 litres/h, temperature 50-98°C, were planned but no operations undertaken. Somehow the exploration of GTE has gone into low priority on account of its high cost of production in comparison to the coal and hydroelectric power constructions.

Environmental Impacts

- A definite impact on the soil environment was observed wherever geothermal power plants were erected. Since the fluids arising from the earth with hot water contain large quantities of compounds, namely hydrogen sulphide, sulphur dioxide, ammonia, boron and sometimes methane, spillage of these on the ground cause pollution. The nearby cattle and agriculture would be seriously affected.

- Power plants of the geothermal energy type are clumsy and damage the scenic beauty of the zone and hence public approval is difficult to obtain.

Questions

1. What is geothermal energy and how is it identified first?

2. Discuss the various methods employed for extracting the geothermal heat from earth. Explain the limitations of effectively bringing all the heat from GT source to the power plant.

3. Discuss the Flash cycle and Binary cycle types of extracting the GT energy.

4. What are the prominent uses of GT energy and which country is adapting this application?

10

BIOMASS

Biomass refers to the non-fossil organic matter that has an intrinsic chemical energy content in it. These are mainly animal waste, wood, seaweed, hyasynth, algae, peel of fruit and vegetables, straw, husk, molasses, bagasse and municipal waste etc. These are all the materials in which solar energy was stored in the chemical bonds of the type C-C.

Photosynthesis is the process through which the carbons are fixed in the biomass. In this process, the plant matter absorbs the solar radiation, fixes the carbon and stores it in its chemical bonds. The equation for photosynthesis is written as follows:

$$CO_2 + 2H_2O + Chlorophyl \rightarrow (CH_2O) + O_2$$

The carbohydrate represented by the building block (CH_2O) is the primary organic product in the reaction. For a gram-mole of carbon to be fixed by solar radiation, the incident light should be in the range of 8-15 % and it should provide an energy of 470 KJ (112 kcals).

The inherent energy from this matter can be recovered through cleavage of these C-C bonds. The extracted energy comes out in the form of heat, and sometimes energy generating substances such as methane, methanol, ethanol or hydrogen results in. During combustion especially, the carbons get converted to the carbon dioxide; and this quantity of carbon dioxide is to the same extent as that which participated in the photosynthesis when the biomass was actually evolved. Therefore, the liberation of energy from biomass do not result in extra greenhouse gas. Actually the use of biomass by the society is the reverse phenomena of photosynthesis. Biomass is the most effective form of stored solar energy. The chief categories of organic compounds present in any biomass are cellulose, starch, lignin, carbohydrates and a few hetero compounds. The average energy content of a few biomass types is shown in Table 10.1.

Table 10.1 Energy content of some biomass materials

Type of biomass	GJ/ton	GJ/m³
Wood, air dried, 20% moisture	15	10
Paper, stacked newspaper	17	9
Dung, dried	16	4
Straw, baled	14	1.4
Baggasse of sugar cane	7-8.8	---
Domestic refuse, as collected	9	1.5
Commercial wastes	16	---
Grass, fresh cut	4	3
Petroleum*	42	34
Methane/propane/butane	50-56	---
Natural gas*, supply line pressure	55	0.04

* chosen for comparison.

Vital statistics on the annual production of biomass in our country are shown in Table 10.2. From the figures recorded here, it is evident that 20,000 MW/year of energy can be obtained. The installed capacity in 40 generation units has been about 223 MW during the year 2001. This portion of energy actually corresponds to 1-3% of the same energy obtained from conventional fossil fuels.

Table 10.2 Annual Production and Utilisation of Biomass in India

Type	Available Quantity in million tons#	Utilisation in million tonnes	Energy Output potential	installed
Wood, for fuel	Approx, 30 times that of Utilisation figures	200- 300	19,500 MW	308 MW
Animal waste	- do -	80 - 100		
Crop residues	- do -	100 - 120		

Hence biomass cannot replace the dependence of the nation on fossil fuels, but can augment to some extent. This is often noticed at locations where the biomass actually originate and is plenty actually e.g., in rural areas. For example, molasses is utilised to produce ethanol at sugar factories situated close to the sugar cane fields in rural areas. Bagasse is burnt in heaters to achieve evaporation of cane juice. Gobar gas, a fuel, is produced from agricultural wastes in rural areas. Almost 50% of wood coming from wayside trees is burnt by village folk only.

BIOMASS CONVERSION

Essentially two methods are applied for converting the biomass into energy directly or through energy-generating substances. These are:

Thermo-Chemical Methods

Thermo-chemical methods

Biochemical methods

Combustion

Direct combustion of some materials of plant origin, namely, wood, straw, husk, bagasse etc., produce sufficient heat to generate steam. By diverting the steam into a turbine, electricity can be generated. Also, the heat obtained during combustion can be used in rural industries such as manufacture of jaggery, jam, syrup etc. Dried cake of cow dung is another fuel of value in villages. The average energy content of some biomass types are shown in Table 10.1.

Social Forestry Yielded Biomass: The wood of certain species such as willow, poplar, cassava, eucalyptus etc., yield tremendous heat and at the same time brings a reduction in the greenhouse gas, namely the carbon dioxide. The biomass from these materials can be successfully utilised for energy development.

Energy Plantations

By definition, *cultivation of any type of plants that store enormous solar energy within and thereby possess high fuel value is termed as energy plantation.* These cultivation and energy extraction processes are managed on an economic scale, so that the plant stands at par with the traditional fossil fuels in the energy-delivery process. Most common species of plantations are from:

Eucalyptus, Casuarina, Terminalia, Leucaena Sargassum (seaweed), Water Hyasynth, Acavia etc.

The following thermochemical processes are adapted to extract the energy from these materials.

Pyrolysis

Pyrolysis or destructive distillation is another form of thermochemical conversion. In this process one type of biomass is converted into a better type of fuel. The operation involves heating the organic material to high temperatures ranging between 300-800 °C distilling out the volatiles and separating the residue. Both residue and volatiles have fuel value and therefore are collected separately by suitable methods. The volatiles are rich in hydrocarbons and appear 'oil-like'. This pyrolitic or bio-oil can be refined and turned into a variety of useful products. The residue is commonly known as 'charcoal' and its calorific value largely depends on the type of original raw material employed. From 4-10 tons of wood , about 1 ton of charcoal can be obtained through pyrolysis. Pyrolysis of wood turnings or saw dust in retorts yield volatile chemicals such as methanol and formaldehyde. Methanol in turn is a fuel that has an energy potential of 23 GJ/ton. It is a direct substitute for gasoline; compared to the fuel-wood, use of methanol in industry or transport as a fuel is more profitable.

Gasification

When a solid biomass of organic nature is exposed to hot steam (H_2O) in presence of excessive air or oxygen at high temperature and pressure, several transformations take place resulting in the formation of gaseous fuels, usually a mixture of CO, H_2, CH_4, together with small quantities of CO_2 and N_2. The process is known as 'gasification'. Gasification of biomass helps in obtaining a much cleaner fuel than the original complex mixture. Also the unnecessary solid particles are avoided in the product formation. The gas can either be burnt directlly to generate heat, or led into an internal combustion engine or gas turbine engine to give power. Biomass material of lignite type when heated with superheated steam gets reformed into carbon monoxide, methanol and hydrogen. Some of these chemicals that are generated have fuel applications. Thus production of gaseous fuels through the 'gasification' operation is very versatile.

Biochemical Methods

Enzymes, yeasts and several microorganisms convert the moist biomass at ambient temperatures into liquid and gaseous fuels. The outcome in these processes is usaually the formation of methane, methanol or ethanol together with some aldehydes. Most agricultural wastes, algae, hyasynth etc., yield methane, which is also called 'gobar gas'.

Production of Gobar Gas

A typical gobar gas plant is shown in Figure 10.1. The tank is divided into two halves by means of a block; one half is the slurry feed stock while the second half is the spent slurry. After the slurry is introduced into the pit it is fed into the tank. The microorganism is also introduced into the slurry, when the biochemical change commences. The biogas forms and gets collected at the top layer of the slurry. The spent slurry flows from the upper layer into other divided half of the tank and exits. The gas is collected from the outlet at the top of the dome.

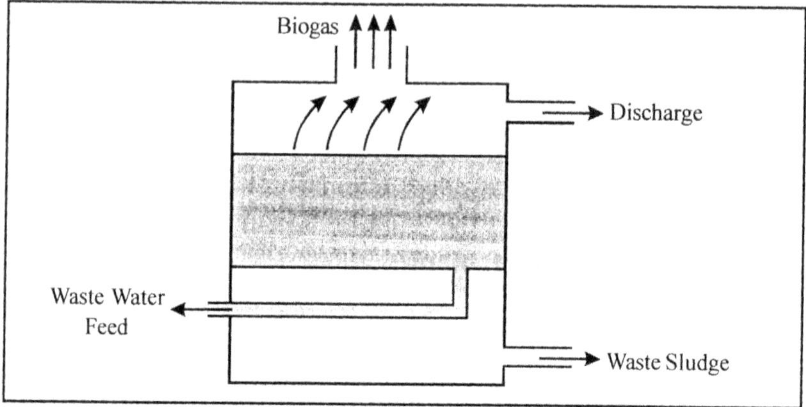

Fig. 10.1 *Line Drawing of BIOGAS Plant*

The size of the digester is uaually 1 cu.meter for small household and 2000 cu.meters for commercial scale production of biogas. The charging of the slurry may be continuous or batch-wise and the digestion process goes on from one day to about ten days. Heat is always generated in the bacterial action and in warm places this is sufficient to push the anerobic reaction forward; but in colder places additional heating is required to maintain the digesters at a minimum temperature of 35 °C. An efficient digestor would produce 200-400 cu.meters of biogas per ton of biomass with methane content ranging from 50-75% i.e., about 11GJ of available energy.

The slurry is made from crop waste, kitchen turnings, cattle dung and faeces. Methanobacterium and methanococcus spp. are the microorganisms used for the production of methane specifically. The steps involved are:

- solubilisation and hydrolysis of organic species,
- acidogenesis,
- methanogenesis.

Biogas

Biogas, comprising of methane, about 60% and the remaining carbon dioxide chiefly together with minor quantities of nitrogen and hydrogen sulphide is produced by digestion of animal/human/plant wastes. It is a clean mixture of combustible gas, easy to handle, not cumbersome as the biomass from which it has originated and economic to produce near the zones of abundance of biomass.

The biomass is placed in the designated container, called the digester and the 'methane fermenter' (the microorganism that is capable of generating methane under anaerobic conditions) in calculated quantity is added; the reactor temperature is maintained between 35-38 °C. Essentially the anaerobic fermentation proceeds through the stages

1. enzymatic hydrolysis of starch, protein and fatty materials yielding broken simple compounds,

2. acidic formation owing to the breaking of organics by 'acid formers' and

3. methane generation due to anaerobic digestion with 'methane fermentors'

All these successive reactions take place simultaneously, the first and second steps taking 24 hours each and the third step about 14 days. The factors that influence the digestion mainly are: pH, temperature, total solid content of the feed matter, its loading rate, seeding of microorganism, control of acid accumulation etc.

Most favourable temperatures for digestion are between 35-38; biogas production below 10 °C do not occur and the optimum thermophilic temperature is around 55 °C. The necessary process conditions vary from the type of digestion, variety of feed stock and carbon-nitrogen ratio of the input feed stock.

Although several types of continuous plants and batch plants are available, the former type carry many advantages in respect of small sized digestion chamber, simplicity, ease of operation, low period for digestion, and continuous gas supply. At least 20 brands of biogas plants are available from agencies such as,KVIC (khadi and village industries corporation), Pragati (social dev. And research program, Pune), ASTRA (application of science and technology in rural areas), designs made by State Governments of Tamilnadu, Punjab, Himachal Pradesh, Roorkee,

Kutch, Gujerat and the like. In addition several pilot plant designs from R&D Organisations are also found in literature and the entrepreneur has to make a proper selection depending on the inputs of feed stock and output biogas expected from the chosen process. Biogas production is handled in a family-scale as well as community scale in rural India. The agricultural wastes, animal debris, water, and other requisites are available in the vicinity of crop area and so this biogas venture is encouraged in these areas. In India, technical consultancy in the establishment of these plants is offered by MNES, CSIR<NRDC and various agricultural universities.

Recovery of Biogas From Landfill : The municipal authorities dump several tons of garbage in landfills earmarked for disposal. Microbiological degradation occurs in these sites in the absence of oxygen and the resultant gas, which is identified as methane oozes out at the surface in cracks.

A landfill site is developed with the complex household /municipal solid waste dumped in a ditch. A lay-out of perforated pipes (interconnected) array is laid upto a depth of 20 meters in the refuse. On the top of the landfill an impervious layer of clay or similar material is built up and this will serve as a cap for the whole municipal solid wastes to decompose into gas through anerobic digestion. Under these conditions at a good landfill site, at least 200-350 cu.meters of gas per ton of refuse can be generated over several years (unlike in gobar gas digesters). The gas is termed as 'landfill gas' (LFG) and this is a mixture of methane and carbon dioxide. The pipes collecting the gas runs several miles in large landfills.

Theoretically each ton of refuse at a site in its life-span of 10 years can produce 200-300 cu.meters of gas that is equivalent in energy to 5-6 GJ. When the yield is 150 cu.meters/ton, total electricity of over 5TWh can be obtained. The economics of power generation from landfill gases are very attractive and countries such as UK and USA have harnessed this energy to the fullest extent. The largest power plant based on landfill gas is operating at the Puente Hills, Ca, USA and is producing 46 MW of electricity. Worldwide there have been 240 power plants operating.

One application of the gas is in the gas turbine engines as a fuel; a mixture of 80 : 20 of methane- diesel runs a GT engine. This power can also be used for running a water pump or lighting the premises or domestic cooking of food. Landfill gas plants making use of segregated domestic wastes under more controlled conditions are also operating in big cities . An advantage of these digesters over the large landfill plants is that they can be installed within the municipal areas thus reducing the transport costs. One such installation in Manchester city, UK, (the Appley Bridge Landfill site of Generators) produce electricity within the municipal limits under controlled conditions.

Fermentation Processes

When yeasts are used to disintegrate the crop waste, plant trash etc., in a fermentor, the biomass gets converted into mixtures of alcohols, usually methanol and ethanol. Of the various processes, those that yield ethanol are preferable, as this substance is not toxic. The food crops that afford ethanol are:

- Sugar materials, e.g., cane, molasses and sorghum.
- Starch materials, e.g., corn, potato, cassava.

The ligno-cellulosic crops of the types –soft wood, hard wood and straw yield ethanol upon fermentation. During the present times biotechnology has provided many improved yeasts that are alcohol-tolerant and yield ethanol to an extent of 12-14%. One such yeast is ***Zymomonus mobilis*** that has many live cells to promote fermentation.

Fermentation Process

The first step in this process is the extraction of sugars from the biomass by crushing and homogenisation in large quantity of water in a tank (in fermenter). Later, a calculated amount of yeast is added and the temperature of the fermenter is maintained at 20°C for 24 hours. The yeast breaks up the sugar into ethanol; the concentration of the alcohol in the mixture is generally between 10-15% and distillation of this mixture in retorts afford the alcohol of 95% purity. Absolute alcohol (99% purity) can be prepared from this product through azeotropic distillation using benzene. Ethanol is usually blended with petrol upto 10% to afford 'gashol'

Approximate yields of ethanol recoverable from several carbohydrate-possessing crops through the above technique are shown below:

Raw Material	Yield of Ethanol Litres/ton
Sugar Cane	50-80
Beet	90-100
Maize	360-400
Wheat	370-420
Barley	310-350
Sorghum	330-370
Potatoes	100-120
Sweet potatoes	140-170
Cassava	170-190
Softwood	190-270
Hardwood	160-230
Straw	140-180

FEATURES

Energy from biomass is low-capital intensive and economically feasible. The complete technology is available in India. Rural areas will be benefited if these are fully adopted. The energy from biomass is renewable and this augments the regular energy in day to day life. The processes are ecologically safe and no air or water pollution is envisaged. Although there is uncertainty in the procurement of biomass in cities and rural side, with adequate planning , the schemes can be pushed with success. Since large area of land is occupied by the biomass and enormous water also consumed in the operations, rural areas are more suitable. In general, the advantages overweigh the disadvantages. The technology is currently adopted in countries, India, China, Brazil and Korea.

The quality of biomass is dependent on the organic material present in it. Generally municipal wastes contain about 50-60% of organic matter. Most of the Indian cities generate about 2000 tons of waste a day. Almost half of this is industrial waste comprising of metal, plastic, rubber, automobile trash, glass and cement rubbish. From the remaining half, the city planners can think of extracting energy. Thus it is an economic proposition to generate energy from at least 1000 tons of organic garbage for town heating services, as each kg of garbage affords 9.5 MJ of heat. Side by side electricity also can be generated from the wastes. In fact, a 10 Mw power plant was erected at Jalkhari in Punjab that works on the heat produced from burning straw. Similar power plants are operating in Andhra Pradesh utilising the rice husk as fuel. About 37.5 MW of power for 24 hours can be extracted from the agricultural waste. The annual output of biomass and the energy content of garbage collected from a few Indian cities calculated on these lines is depicted in Table 10.3.

Table 10.3 *Average Content of Solid Wastes produced in Indian cities and available energy (1990)*

Indian city	Content of Solid Waste (tons / day)	Calculated Energy GJ/day
Mumbai	4260	19574
Kolkata	3900	17920
Delhi	3200	14704
Chennai	1522	12144

Source: RC. Yadhav, 1995.

BIOMASS MATERIALS AND THEIR TYPICAL CHARACTERISTICS

Biomass materials and biomass based industry residues are very important renewable energy sources. The importance of these materials as alternate fuels has been well recognized for use in power generation. While the characteristics of biomass vary from different geographical regions, the values for typical biomass materials are indicated for general guidance.

Biomass	Grade	Bulk Density kg/m³	Ash Content %	C %	H %	N %	O %	Calorific Value Kcal/Kg.
Castor stick		5.40	45.97	6.65	1.28	40.70	4300	
Castor Seed Shell		8.00	44.25	5.65	0.16	41.94	3860	
Corn cobs	11% moisture	304	1.20	41.44	5.96	0.14	51.26	
Cotton pods			5.01	41.49	6.20	1.81	45.49	4200
Cotton stalk			3.01	41.49	6.20	1.81	47.49	4200
Saw dust	Loose	177	1.20	52.28	5.20	0.47	40.85	4400
Straw	Loose	80	15.50	35.97	5.28	0.17	43.08	3700
Straw	Bales	320						
Wood (hard)	—-	330	1.50	52.30	5.20	0.50	42.00	4400
Bagasse	0.12	74	4.00	47.00	6.50	0.0	42.50	4200
Coir pith	0.12	47	13.60	41.27	4.02	1.51	39.60	4100
Cotton shell	0.21	79	4.60	44.19	5.87	0.73	44.61	4200
Coconut Wastes			6.31	46.69	5.89	0.07	41.04	3720
Coffee Husk			11.61	46.46	6.26	0.72	34.95	3745
Eucalyptus Saw dust	0.12	239	0.21	49.37	6.39	2.02	42.01	4400
Ground nut shell	0.15	165	3.10	33.90	1.97	1.10	59.93	4500
Rice husk	0.12	235	22.20	36.42	4.91	0.59	35.88	3200
Sun flower stalk	0.12	93	4.30	44.20	5.50	0.50	45.50	4300
Sugar cane leaves	0.13	167	7.71	39.75	5.55	0.17	46.82	4200
Saw dust	0.12	165	1.20	52.28	5.20	0.47	40.85	4500
Subabul			1.20	42.76	5.68	1.07	49.29	3980
Sweet sorghum stalk			7.40	41.83	5.90	0.53	44.34	4100
Tobacco dust	0.18	586	49.40	19.94	1.63	4.35	24.68	1164
Tea waste	0.17	100	6.70	45.97	5.40	2.76	39.17	4000
Tobacco stem			20.6	28.30	4.50	1.00	45.60	3041

PLANNING AN ENERGY GENERATION PROJECT UTILISING CITY GARBAGE

Most of the cities get municipal wastes from the residences, offices, hospitals, schools, shopping and markets, public services such as railway, bus, waterways, dairy and processing and heavy industry.The waste developed at each of the locations is specific and it is not possible to develop a common evaluation procedure. Moreover, the waste from some of these sites may not always be adding up to municipal pool, as the individual units dispose off at their sites following other techniques. However, the general methodology to be adapted shall be as follows:

Name of the location and activity:
Nature of work:
No. of persons at work:
No. of hours activity:
Which materials are dealt:
Identification of waste:
Quantity of the waste, category -wise:
Whether waste/ garbage is segregated at first despatch:
Whether any methods adapted for

- reduction
- reuse
- recycling any technology adapted for producing energy:
- Any modifications/ R&D available:
- Any pilot plant activity going on:
- Major plant erection:
- Economy gained:
- Control of environmental pollution:
- Characterisation of the waste (physico-chemical)*

Domestic residence, dairy, poultry, market, shopping, hospital, school, industry such as, food manufacture, chemical, fertiliser, cement, chemical, agro-based, railways, port, bus station, office etc.

Physico-chemical characteristics of wastes

After segregating the waste for organic and other type, the following data has to be obtained experimentally.

Appearance and texture:
Whether solid or massy:

Density

A box of volume 1 cu.m filled with the waste and weight measured.

Proximate Analysis

- Moisture - loss of weight at 105°C within 1 hour.
- Volatile matter- loss at 950°C
- Ash- residue after burning

Calorific value

Ultimate Analysis:

- Determination of elements, C, H, O, N, S.
- Estimation of metals.
- BOD
- Microbiological examination.

  Energy content of solid waste (empirical)

  $$E = Btu/lb = 145.4\ C + 620\ (H + 1/8\ O) + 41\ S.$$

  $$E\ (on\ dry\ basis) = Btu/lb\ above\ (100\ /\ (100 - \%\ moisture)$$

Questions

1. How is biomass formed on the earth?

2. Explain the sequence:

   Solar energy——Photosynthesis————Biomass——— Energy.

3. What is meant by 'anaerobic digestion' ? what are the factors that influence this process?

4. Give a sketch of a 'Gobar gas' production plant and explain its functioning.

5. How is 'landfill gas' obtained? What is its composition? How is this gas used?

6. Explain the process of wet digestion of biomass in digestors and give the composition of biogas.

7. Write notes on wet fermentation to obtain ethanol from biomass. What are the advantages in obtaining products of this nature?

8. Write notes on energy plantations.

11

BATTERIES AND ELECTRIC POWER

Batteries are electrochemical systems that can deliver electric power at the desired time. The chemical reactions that take place between the ions in an electrolyte and the electrodes kept in the same solution cause the movement of electrons thereby generating electricity. Thus if two iron rods are placed in a solution of sodium chloride, a potential difference is observed between the two electrodes and this is the electric current that has generated. By definition this system is called an electric cell and a combination of several cells constitute a battery system. A cell contains a positive and negative plate separated by non-electrical porous sheet such as rubber or polymer. The charged plates are of metal, usually lead. A single cell with fresh ingredients in charged state shows a potential difference of 2.0 volts.

As one draws power from these 'combination cells', the voltage of the system falls low. This can be revived once again by passing electric current into the electrolyte and reversing the chemical state of the two electrodes.

The process is known as 'charging'. If a battery system is amenable for 'charging', thereby affording the original power, it is classified as 'secondary battery'. On the contrary if the charging cannot be achieved to restore it to original state, and the drawl of power is only once from the system, it is denoted as 'primary battery'. In fact, both types of batteries find applications in industry. An additional feature of the secondary battery is that it can be used to store excess electricity within the system when it remained in 'discharged state'.

The secondary batteries display:

- High delivered energy in terms of watt-hours per kilogram of system, i.e., Wh/kg,
- Capability of yielding high energy for sustained period,
- High electrochemical power efficiency, and
- Function for long cycles of charge-discharge.

The secondary batteries in common use today are: lead- sulphuric acid type, nickel-iron with alkali type, nickel-cadmium, nickel-zinc, lithium alloy- metal sulphides type etc. Lead-acid batteries are comparatively cheap, have good endurance and are therefore widely in use. A detailed description of various features of the secondary batteries with the example of lead-acid type has been given below:

A lead- acid battery is built up of several small cells (Fig.11.1). The cell plates are made up of lead and the electrolyte is sulphuric acid of specific gravity 1.3. The plates are separated in a cell by means of separators which are porous in nature and electrically non-conducting. The +ve plates are made up of lead core coated with lead peroxide and the - ve plates with lead core coated with sponge lead. The lead core is a hardened lead and antimony (3 % Sb) alloy. The battery is a combination of six cells each of 2 volt capacity in series, and hence offers 12 volts of electric power. In the battery grid, the core lead is coated with either sponge lead or lead- antimony hardened alloy. Several batteries can be linked in series or parallel or both ways depending upon the requirement of current.

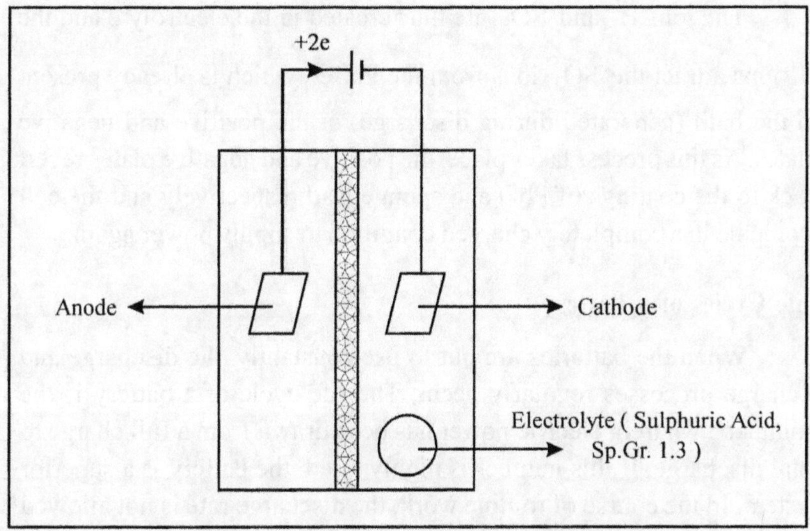

Fig. 11.1 *Line Drawing of Lead Acid Battery*

Discharge Pattern

The electrolyte ionises into H^+ and $\overline{SO}_4$ on drawing the current from the battery. The $\overline{SO}_4$ ions slowly settle at the negative plate, extract lead from sponge coating and becomes white lead sulphate, $PbSO_4$. Similarly the H^+ takes off oxygen from the lead peroxide at the positive plate, forms water, resulting in the formation a 'coated' lead sulphate. Thus at both the plates, the white $PbSO_4$ sets in. This is the discharge pattern of the battery on drawing the current. On account of formation of water, the strength of sulphuric acid falls down below 1.3. Also, the original cell voltage which was at 2.2 diminishes to about 1.7 after the current was drawn.

Recharge Pattern

During the recharging process, the strength of sulphuric acid has to be stepped up to a density of 1.3 again, and the voltage of the cell to be raised from 1.7 to 2.2 by passing a DC supply in the electrolyte and reversing the discharge process.

The ions H^+ and $\bar{S}\bar{O}_4$ are thus created in the electrolyte and the H^+ ions attract the $\bar{S}\bar{O}_4$ ions from the $PbSO_4$ which is already present in the bath (generated during discharge) at the positive and negative plates. As this process takes place, the positive and negative plates revert back to the coatings of PbO and sponge lead respectively and the cell would be in a completely charged condition to supply power again.

Life Cycles of a Battery

When the batteries are put to use constantly, the discharge and recharge processes regularly occur. The life cycle of a battery is the 'number', wherein electric power has been drawn from a full charge to total discharge. If this number is highly rated, the battery is a superior system. In the course of routine work, the discharge rate is not allowed to go beyond 80%, that is, the battery power will not be allowed to fall below 20%. At the same time excessive recharging above 80% of the effective power also is not desirable. Therefore the battery would be in a fit condition between 20-80% of its discharge rate and 20 upto 80% of its recharging rate. In automobiles, the battery is generally not allowed to discharge below 50% as a precaution. It is possible to determine the life cycles of battery in an automobile and assess its performance.

Rating of Battery

The rating is expressed in terms of the current strength inherently stored in the battery. It is expressed in units of ampere- hours (ah). The battery manufacturers denote the rating in 100 ampere hours for comparison with other types. Thus, if a battery is rated as 100 ah at 4 hours, it means the discharge rate is 25 amperes for 4 hours. It can also be 50 amperes for 2 hours or 10 amperes for 10 hours and so on. But in reality at higher rates of discharge, such as 100 amperes for 1 hour or 50 amperes for 2 hours, it will not be exactly as per the equation, and it may result in 100 amperes for 45 minutes or 50 amperes for 1 hour 45 minutes or so. At lower rates of electric discharge, the supply of current is steady.

Battery rating is also dependent on the mass of lead present in it. The more the content of lead in the battery, the more it can store the amperes. A 50 lb battery with a rating of 200 ah discharges faster than a 150 lb one with a rating of the same 200 ah. The electro-chemical energy in a battery is expressed in terms of kilowatt hours, which is calculated as:

$$Kwh = ah\ (rating) \times voltage\ /\ 1000.$$

Thus a battery array of 120 volts with 200 ah rating will have 24 kwh energy in it. If one draws energy from the array at the rate of 4 kwh per day, the energy in the battery system would be sufficient for 6 days of drawl. This calculation will be used when battery power has been used to augment other types on non-conventional energies whose availability is intermittent.

Battery Connections

Interconnectivity of several batteries is made in two ways:

Series Connection

The +ve terminal of one battery is connected to the -ve terminal of the other and so on in series. The total number of batteries are arranged in two rows and this sort of connectivity is made in each row. After the system is developed the first row will have one terminal, say +ve, the other row will have the –ve terminal. This arrangement will increase the total voltage of the battery bank according to the formula:

$$Total\ voltage\ of\ the\ battery\ bank = Voltage(single\ battery) \times total\ number\ of\ batteries\ connected\ in\ series$$

Thus if 20 batteries of each 12 volts are connected in series, the total voltage of the system would be $20 \times 12 = 240$ volts. The ampere-hour of the individual battery would be the same as that of the total system, that is to say that the storage capacity of the battery bank is same as the individual one.

Parallel Connection

In this arrangement the +ve terminal of one battery is connected to the +ve terminal of the other and so on, in order to group a large number of batteries. The result is a parallel connection to all the positive as well as the negative terminals of the individual batteries. The electrical pressure or voltage in the array remains the same while the storage capacity of energy increase in multiple to the number of batteries connected parallel. If 4 numbers of 12 volts (each) batteries are connected in parallel, and if each of the battery is having a rating of 200 ah, then the output voltage of the battery system will be 12 volts only, whereas the storage capacity will be:

$$4 \times 200 = 800 \text{ ah.}$$

Series-Parallel Connection

This is yet another method of connectivity of the battery terminals. The batteries are arranged in one row only, unlike in series connection. The +ve terminal of the first battery is connected to the -ve terminal of the second one. Similarly, the pattern is repeated for the second and third batteries. It goes on for the entire set. The pattern would give a picture of Series-Parallel connectivity. The three patterns are shown in Figure 11.2.

It is very important that the batteries of identical type alone are to be used while making series or parallel connections, as otherwise inequalities in discharge rates set in. Also the batteries shall be trim, maintaining sufficient acid levels and properly mounted. This sort of positioning is particularly important in submarines that maneuver in the seas.

Inspection of Battery

The following checklist has to be followed to maintain the batteries in trim condition:

- Voltage shown by the individual battery. This should be in accordance to the manufacturer's specification.

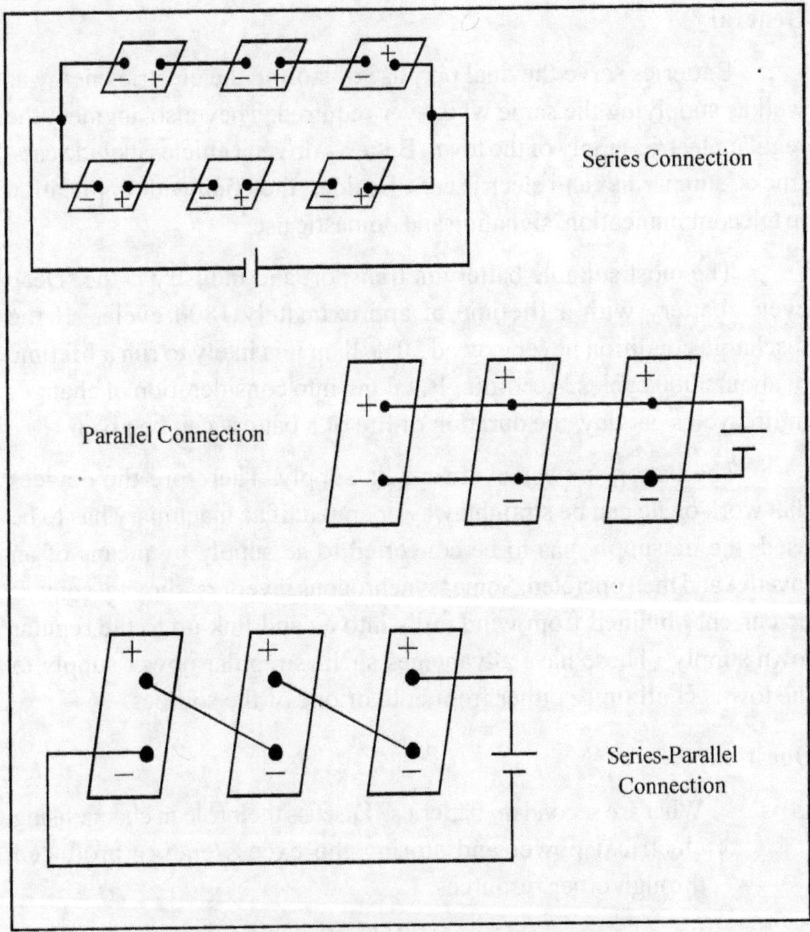

Fig. 11.2 *Inter Connectivity of Batteries*

- Checking the specific gravity of the sulphuric acid, in case of lead- acid battery. In case of alkali batteries, finding the strength of alkali by titrimetric techniques.

- Maintaining the level of the electrolyte in the cell.

- Inspecting the corrosion of surrounding parts due to acid/ alkali and adapting measures to check the spill of the electrolyte.

General

Batteries serve the dual purpose of storing the electric energy as well as supplying the same whenever required. They also augment the regular electric supply of the town. Battery- driven vehicles include cars, trucks, submarines and electric cars. Besides, they find wide application in telecommunication, signaling and domestic use.

The most suitable battery in transport and industry is the 'Deep cycle' battery with a lifetime of approximately 1800 cycles. If the discharge condition never exceed 50%, then it is likely to run a lifetime of about 5000 cycles. Accordingly, taking into consideration of changes in lifecycles per day, the duration of life of a battery can be given.

The battery power is always dc supply. Therefore the devices that work on dc can be straightaway operated; if ac machinary has to be used, the dc supply has to be converted to ac supply by means of an inverter and then operated. Some 'synchronous inverters' directly convert dc current obtained from wind mills into ac and link up to the regular town supply. These have advantages such as regular power supply to the towns at all times either from both or one of the sources.

Questions

1. What are secondary batteries? Discuss their role in channelising electrical power and storing the excess energy produced through other resources.

2. Write notes on Lead-sulphuric acid battery, the inter-connectivity of multiple batteries and inspection maintenance in huge power delivery systems.

12

NUCLEAR ENERGY

The scientific discoveries made during the first three decades of the 20th century conclusively established the existence of positively charged protons, negatively charged electrons and the charge-devoid neutrons. The model of atom conceived by the British scientist, Ernst Rutherford was widely accepted. Later discoveries made by German scientist, Wilhelm Rontgen disclosing the formation of a new invisible radiation in the electromagnetic spectrum on bombardment of matter with high- speed electrons resulted in the X- rays. Many scientists across European countries studied the bombardment of matter with fundamental particles and discovered the phenomenon of radioactivity and radioactive elements. Further, it was discovered that neutrons, possessing no electrical charge, are not deflected during bombardment by any other charged particles and so are capable of penetrating the nucleus of any other atom and destabilise it or cause splitting. Adapting this technique, the Italian scientist, Enrico Fermi found that more than 40 of the existing elements could be obtained in the neutron bombardment nuclear reactions in the radioactive form.

Uranium is one of the elements of this type. During the same period, the German scientists Otto Hahn and Fritz Strassmann as well as the Austrian scientist Lise Meitner discovered the same phenomenon and came up with the idea that a 'nuclear fission' has been operative with the formation of radioactive element barium. They also correctly deduced that the energy released in the 'fission' corresponds to the calculated energy enunciated by Albert Einstein, namely

$$E = mc^2,$$

where E = energy,

              m = mass and

              c = the speed of light.

Thus a very small quantity of matter when subjected to fission can result in huge energy through the disruption of the binding force holding their nuclei. It is this *huge* energy (in the form of heat mostly) that is termed 'nuclear energy'.

Whenever a radioactive substance is split, enormous energy is liberated together with the formation of various other substances. The splitting is achieved by means of the heavy particles, neutrons. Thus Uranium-235 is susceptible for neutron bombardment to ultimately yield lower elements, heat and additional neutrons. This is an example of a nuclear fission reaction. The additional neutrons participate in further reactions and sustain the release of energy. To start with, the neutrons may be within the radioactive material and when brought to a critical stage, the fission develops; alternatively the neutrons from an additional sources may be provided or from within the same nuclear reaction itself.

Nuclear Fuel

The heat generated in a nuclear reaction involving a high density element that is 'fissionable' and fast moving neutrons is the starting point for conversion into electricity. In terms of generation of power, the 'fissionable' matter is called the nuclear fuel and the process , the nuclear fuel cycle. The choicest nuclear fuel so far available is the metal uranium.

It is 1.7 times denser than lead. In its nucleus, 92 protons exist while the number of neutrons change between 140-143 depending on the isotope. The naturally occurring form of uranium is the U^{238} (about 99.3 %); it is accompanied by U^{235} (about 0.7 %). Of these two, U^{235} alone is fissionable and the optimum concentration of this element for a nuclear reaction is about 3-4%; hence, the natural uranium has to be enriched prior to subjecting the same in a reactor.

MINING

Uranium is mined from earth in many countries, e.g., Australia, Kazakhstan, Canada, S.Africa, Brazil, and USA. It occurs in many rocks in concentrations ranging from 2-4 ppm. It coexists along with tin, molybdenum, tungsten and lead in the earth's crust (average 1.4 ppm). It is also one of the minor constituents of seawater. The 'front end approach' of the nuclear fuel cycle is the mining, milling, conversion, enrichment and nuclear fuel fabrication. The 'back end approach' is the recovery and disposal of the nuclear fuel after it was used in the reactor and spent (spent fuel). Necessary technologies are available for undertaking these processes and extreme precautions are to be taken.

Uranium bearing ore is mined from the earth's crust from the surface open-cut or underground-cut techniques, depending on the 'ore body'. The underground mined ore contain small quantities of copper, gold and silver too. The ore is crushed, ground to powder and a water-slurry is made. Subsequently it is leached with sulphuric acid to remove uranium from other silicates and finally precipitated as uranium oxide, U_3O_8. This mass is called 'yellow cake' although it appears in khaki colour. The oxide is the finished product for commerce.

CONVERSION

Uranium oxide, U_3O_8 is first converted to the vaporous uranium fluoride, UF_6. This vapour is pushed through several screens wherein the lighter U^{235} fluorides move first and it is segregated. The process is repeated several times till significant enrichment is achieved. Finally, the U^{235} fluoride is converted back to uranium dioxide UO_2.

FUEL FABRICATION

The uranium dioxide is powdered and shaped into 2/3″ long pellets, each encased into a ceramic shell and then packed into a 12″ zirconium-alloy (zircalloy) or stainless steel rod. Such rods in clusters form the *nuclear fuel.* It is to be noted here that the nuclear fuel is entirely different from the conventional fossil fuels both in structure and properties.

Pu^{239} is produced from non-fissionable U^{238} through a series of nuclear reactions and is also fissionable. Th^{232} produces U^{233} through several nuclear reactions. Thus we have a variety of fissionable radioactive materials to participate in reactions and generate energy. In the process some short-lived isotopes of elements as well as non-radioactive metals of low atomic weight are also encountered but they are not fuels. Some of the nuclear reactions are :

$$U^{235} + n\,1 \rightarrow X + Y + 2.5n + 4.7 \times 10^9 \text{ KCal (energy)}$$
$$\text{Mass of } X = 85 - 110 \text{ and } Y = 125 - 150.$$

Conversions

$$U^{238} + n1 \rightarrow U^{239} \rightarrow Pu^{239*}$$
$$Th^{232} + n1 \rightarrow Th^{233} \rightarrow U^{233*}$$

* denotes fissionable material

NUCLEAR FISSION

Harnessing Nuclear Power : The heat generated during a nuclear reaction is extractable and the energy liberated is utilised for various purposes. As we observe the above reactions, the heat is of the order of 3.2×10^{11} Joules, or 200MeV per atomic weight of U^{235}. Therefore the reaction has to be systematically controlled to the advantage. The reactions are performed in a 'nuclear reactor', which is a pressure vessel. In this the fissionable material namely Uranium is placed. For a reactor with an output of 1000 megawatts (MW), the core would contain at least 75 tons of enriched uranium. The material undergoes bombardment with fast moving neutrons inside the reactor in a controlled manner and gets consumed. This is very much different from the real burning of a fuel, in the sense that no fire comes but heat alone.

The replacement of at least one third of the spent fuel from the reactor during an year's operation maintains high conversion to energy. This heat is exchanged by the coolant, usually water, that turns into steam automatically. The steam drives the turbine thereby generating electricity. Essentially, nuclear reactions in a reactor provide energy for running a thermal power plant. The basic requirements for nuclear energy-based thermal power generation are:

- Fissionable materials e.g., U^{235}, U^{233}, Pu^{239} etc. In case U^{235} is used as the fuel, it should be in the enriched form namely 2-3 % instead of the natural 0.7 %.

- A moderator that moderates the speed of neutrons to the optimum, eg. heavy water

- A controller that mops up the excessive bombardment of the neutrons thereby permitting a facile chain reaction. The build-up of this system is through the placing of cadmium or boron rods at suitable positions. If excessive chain reactions are not arrested, the heat that comes in the reactor would damage the body of the reactor.

- A coolant that takes up the heat generated in the controlled nuclear reaction. The coolant protects the reactor by lowering the interior temperature. Common coolants are the liquids such as water or heavy water, gases of the type carbon dioxide, air as well as solid sodium metal.

- Finally a shield that do not permit the radiation produced inside the reactor to reach outside. Lead is the shielding material.

Nuclear reactors are of two types, both providing heat (energy) which was harnessed into electricity. These are:

BOILING WATER REACTOR

In this system the pressure vessel of the reactor has an inlet of water. Water enters the pressure vessel and receives the excessive heat generated during the nuclear reaction, gets converted into steam and pushes out through the stem and valves at high pressure of about 170 kg/ sq.cm.

The steam is further led into the turbine to produce electric power. After driving the turbine the steam gets condensed and re-enters the pressure vessel after slight treatment. A line drawing of a Boiling Water Reactor is shown in Figure 12.1.

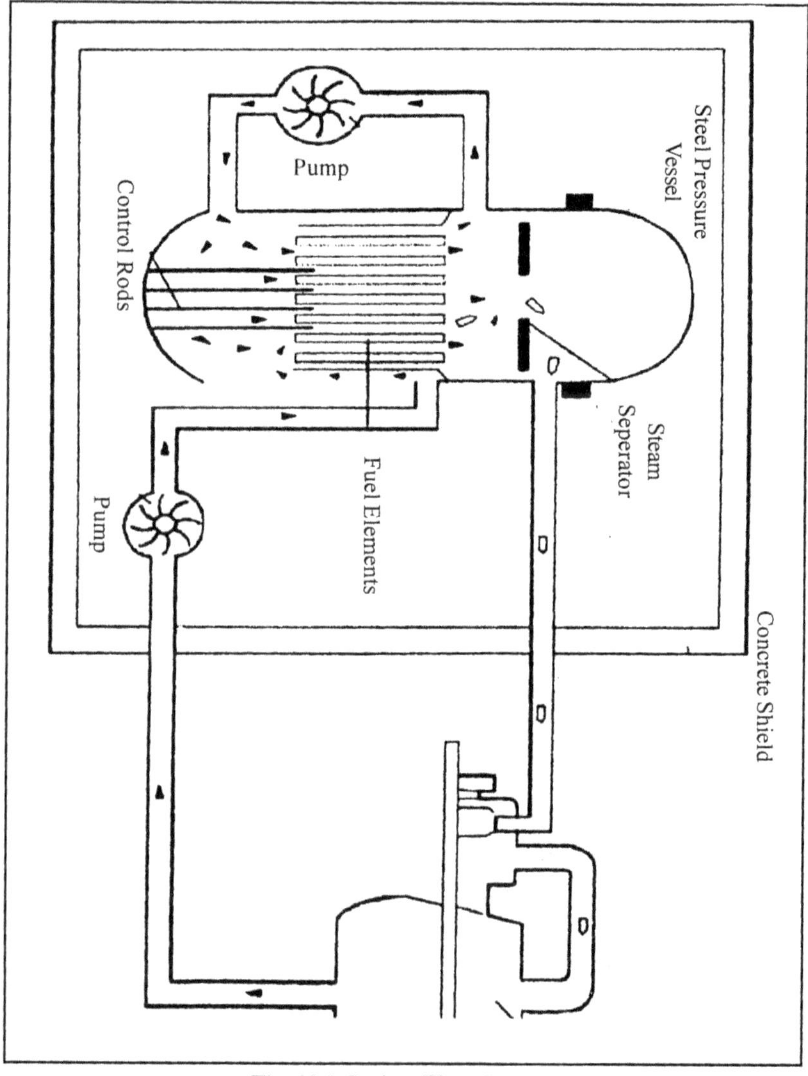

Fig. 12.1 Boiling Water Reactor

PRESSURISED WATER REACTOR

A schematic representation of the pressurised water reactor is shown in Figure 12.2.

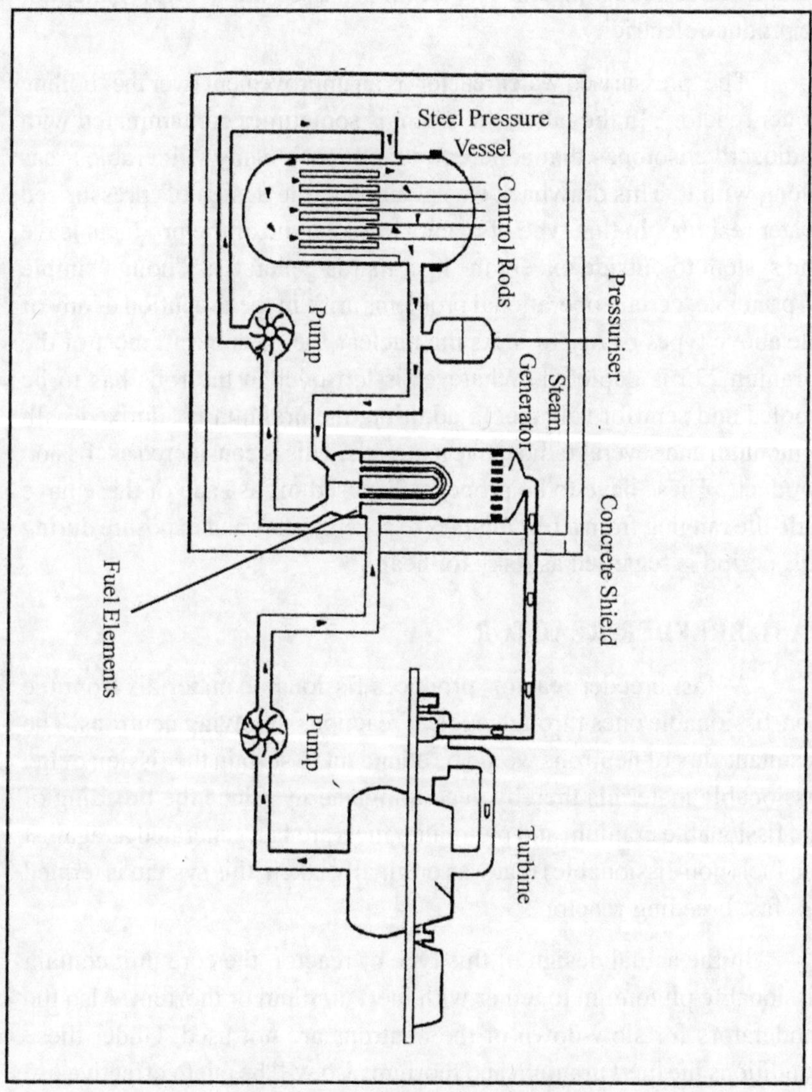

Fig. 12.2 Pressurised Water Reactor

Within the core of the pressure vessel, water at high pressure and a temperature of about 325°C is produced from receiving the heat developed in the fission reaction. This water enters the ' loop' and exchanges its heat with surrounding water and convert it into pressurised steam, about 400kg/ sq.cm. The steam is passed into a turbo-generator to produce electricity.

The 'pressurised water reactor' is an improvement over the 'boiling water reactor'. In the latter, the steam is sometimes contaminated with radioactive isotopes that adhere to it and reach many vulnerable areas along with it. This drawback was avoided in the design of 'pressurised water reactor'. In this type of reactor vessel none of the products leave the system to outside except the heat that is generated. Though simple in principle, certain operational problems arise in the execution of any of the above types of reactors. As the nuclear fuel gets burnt, most of the Uranium235 is depleted. Whatever is left over in the rods has to be pooled and sent for salvage. In addition, the uranium is admixed with plutonium and several radioactive materials. This is considered as 'fission product'. These have to be properly disposed off as each of these have half-life ranging from a few minutes to several years and exposure during this period is regarded as risky for health.

FAST BREEDER REACTOR

A 'fast breeder reactor' produces fissionable materials from the non-fissionable ones through nuclear reactions involving neutrons. The resultant flux of neutrons would be abundant to sustain the fission of the fissionable materials thereby generating energy. Since the breeding of the fissionable uranium and plutonium fuels are fast generated as against the fuel (non-fissionable) that was originally taken, this system is termed as 'fast breeding reactor'.

In the actual design of this type of reactor, the core unit contain fissionable plutonium together with inert uranium or thorium. Also the moderators for slow-down of the neutrons are not used. Under these conditions the inert uranium and thorium also will be put to effective use to generate energy.

The core of the pressure vessel is also small and the exchange of heat is made with the contact of metallic sodium, which do not interfere with the speed of the fast neutrons. Efforts are still put up in advanced countries to improve the designs and applications of the fast breeder reactors.

NUCLEAR FUSION

The researches in many advanced laboratories throughout the world during the years 1970-1980 led to the discovery of the "nuclear fusion" phenomenon. In Nature, several nuclear fusion reactions occur constantly around the Sun wherein the hydrogen nuclei are converted to helium together with the generation of radiation. It was stated that about 650 million tons of hydrogen undergoes fusion at the surroundings of the Sun every second. The intermediary forms of hydrogen, namely deuterium and tritium are also formed in these fusion reactions. It appears one out of every 500 hydrogen atoms found in Nature is deuterium. Within the laboratory, procedures for separation of deuterium have been perfected and a good deal of research using this isotope in controlled fusion experiments has been successfully carried out.

Nuclear fusion occurs between two nuclei of light elements e.g., deuterium (D_2) and tritium (T_3) at a very high plasma temperature of the order of $10^8\,^{\circ}C$ in a reactor resulting in the formation of helium, another light isotope and enormous heat. It is therefore a thermonuclear reaction in which heat of the order of some GW is produced and this energy can be harnessed towards the production of steam. At high temperatures, namely $10^8\,^{\circ}C$ the matter is usually converted to a plasma; it is a fourth state of matter, after solid, liquid and gas, in which the electrons are stripped off from their parent nuclei and the matter remains as a region of highly ionised electrically charged particles. Such a plasma in the fourth state of matter, if in contact with the reaction vessel, can melt the reactor body. It is therefore contained in a compact space within a vacuum chamber under the influence of a 'magnetic field'. This arrangement is known as 'Tokamak Magnetic Confinement' (closed Toroid). A few other techniques are also known to hold the plasma without touching the reactor body.

The following technical particulars of the Tokamak reveal the characteristics of the plasma that generates in a typical nuclear fusion reaction:

Parameter	Measuring Unit
major radius of Torus	7.00 m
plasma radius	1.94 m
ion density	0.81×1014 ions/cm^3
ion temperature	11 kev
max.Toroid field	5.2 Tesla
plasma current	10 mega amp
fusion power density	0.7 MW/m
fusion power output	4000 MW
electrical power output	1200 MW
heating	radio frequency
tritium burnup	42%
burn time	continuous
wall material	special stainless steel
coolant	pressurised water
power cycle	steam/water
breeder	$LiAlO_2$

Source: S.Rao and DB.Parulekar, "Energy Technology", 1999, pp.654-655.

The reactions are represented as follows:

$$D + T \rightarrow He(+3.5Mev) + n (+ 14.1 \text{ Mev}) \text{ (exothermic)}$$
$$Li_4 + n \rightarrow T + He_6$$
$$Li_4 + n \rightarrow T + He_7 + n$$

During the first phase of reaction, a total of 17.6 Mev of energy is generated; in the second phase, the neutrons released in the first phase react with the lithium contained in the mantle of the pressure vessel and produce the requisite tritium, which goes again in the first phase reaction.

As we notice the above reactions, about 17.6 Mev energy is released out of one fusion. The plasma (heat) is so hot that it would even damage the walls of reactor even and therefore it is suspended in vacuum in a device with the help of magnetic field in such a manner that no contact with the body of the reactor is established. The energy, however, can be drawn when required using sophisticated devices. Since deuterium is available plenty in Nature, nuclear fusion reactions can take place continuously and this type of energy is renewable. However the technologies in respect of producing on a commercial scale are yet to be perfected.

Since the basic materials involved in fusion reactions are hydrogen, its isotopes and helium only, no new isotopes of higher elements are envisaged. Automatically, there would not be any nuclear waste formed. As the plasma is always stored in suspension there is no fear of the reactor chamber getting damaged. On account of these it may be stated that fusion energy is safer than fission energy..

Disposal of Waste: Immediately after the nuclear reaction progressed in the pressure vessel, the fuel rods were burnt and several waste materials are produced. These are radioactive with half-life ranging from a few minutes to years. Since these materials are not at all useful for dealing in further reactions, they are termed 'spent fuels'. The spent fuel is first dumped into special ponds and temporarily stored. Spent fuel still contains 96% of U^{235} and therefore it is reprocessed by adapting several metallurgical processes and the final waste is sent for disposal. The disposal procedures are different for low-level wastes eg. from small industries, laboratories or hospitals; intermediate -level wastes arising from industries. Both these types are disposed off by burial underground and only in case of high-level waste from nuclear power plants, it is sent for vitrification. In the thermo-nuclear plants low and medium grade atomic wastes are formed. These have short half life and can be carefully encapsulated in repositories. The process involves the separation of the actual radioactive material from the bulk waste by adopting physical and chemical methods such as filtration, evaporation, chromatography, reverse osmosis etc. followed by immobilisation in matrices such as cement, bitumen or polymer.

These are packed in sealed containers and then buried in reinforced concrete structures down below the earth. The environment around these repository is frequently monitored by drawing soil, water and air samples for radioactivity. During the reprocessing of spent uranium or plutonium rods in the reactors, high grade wastes whose radioactivity persists for many years are generated. These are environment-risky and therefore are encapsulated in inert glass and then packed in special corrosion-resistant metallic containers and buried in granite soils deep below. On account of low water availability in the location, the waste transforms in normal way over a period of time without being subjected to undue corrosion and rapid degradation

NUCLEAR REACTOR ACCIDENTS

In spite of extreme care taken in the selection of reactor materials and designs, proper operation documents and good management by personnel, reactor accidents in cases of ships as well as ground stations have taken place. Lee Davis in her book entitled 'Environmental Disasters' discussed several nuclear disasters that have taken place on the entire globe. Nuclear disasters at the Three Mile Island Pa, USA (1979) and Chernobyl in USSR (1986) figured most prominent. The Chernobyl accident killed about 36 persons on the spot and caused exposure of radiation to approximately 2500 people. The radioactive material that got dispersed to the extent of around 7000 kg possessed a strength of about 100 million Curies. It consisted of radioactive iodine-131, strontium-90, caesium-137 and plutonium. The dust has fallen on many populations, foodstuffs and agricultural produce in the nearby Scandinavian countries and European continent. As a result of this the public and the harvest was severely affected. Although the symptoms of exposure were not causing immediate damage, it was feared that this is a serious ailment. Many advanced countries rendered cooperation in combating the environmental damage.

Nuclear disasters that take place in ships create some logistic problems too. The reactor cannot be left in sea water or at shore. If it is due to wreckage of ship, the damaged reactor has to be salvaged first. In the case of 'Kursk submarine' accident, this drill was followed. A list of accidents involving nuclear reactors is given in Table 12.1.

Table 12.1 Major nuclear power plant accidents

Name of Power Plant	Power capacity	Year	Environment damage	Remedies
NRX Canada, exptl.	40 MW	1952	nil	Repaired; closed 1992
Windscale-I UK, pile	---	1957	Widespread contamination, 1.5×1015 Bq	Entombed; being demolished
SL-1 USA, exptl.	3 MW	1961	Minor release	decommissioned
Fermi-1 USA, exptl breeder	66 MW	1966	nil	Repaired; restarted 1972
Lucens Switzerland, exptl.	7.5 MW	1969	Minor release	decommissioned
Browns ferry USA, commercial	1080 MW	1975	nil	repaired
Three Mile Island-2 USA, commercial	880 MW	1979	Minor short term radiation dose; delayed release of 2×1014 Bq	decommissioned
St. Laurent-A2 France, commercial	450 MW	1980	Minor release; 8×1010 Bq	Repaired; decommissioned; 1992
Chernobyl-4 Ukraine, commercial	950 MW	1986	Major release across Europe; 11×1018 Bq	entombed
Vandellos-1 Spain, commercial	480 MW	1989	nil	decommissioned

Environmental Damage

Exposure to radiation causes several biological defects in human beings and this is the primary damage caused by nuclear reactors. To day we have over 400 nuclear reactors functioning in the world, of which about 120 were in USA. The number of service breakdowns and emission of slight radiations count upto some 3600 between the period 1970-1990.

Dumping of radioactive waste in the underground or oceans also create environmental pollution. Laying the low-level wastes in landfills should be discontinued as these environments are out-of-sight and do not get monitored frequently.

Reactor meltdown, which is the chief cause of many accidents has to be attended immediately, as otherwise the atmosphere, water and soil could be contaminated with radioactive wastes. This is the main reason why many advanced countries have stopped erection of power reactors since 1985. On the whole nuclear power technology is not an eco-friendly technology.

FUTURE

Nuclear power generation throughout the world received a great set-back in present times. There are many who advocate the necessity of producing more nuclear power and as many who oppose. Some of the reasons for zero growth of this energy are as follows:

- The establishment of power plants/ reactors in ships etc., is cost-intensive.

- Most of the nuclear power generation units in the world are running only to about 60% utilisation.

- The minerals for the supply of nuclear fuel in the world are expected to serve only for another 50 years, when the rated efficient working of a power plant is given as 40 years.

- The technology for erection of power plants is not available in many developing/ under developed countries.

- Nuclear disasters are unwelcome; however much care is taken, some failures do happen and the human loss as well as unsettlement occurs.

- The technology for combating nuclear disasters is limited to advanced countries only; other nations have to face the penalties of losses.

- The cost of electric power is just comparable to that obtainable from other non-renewable resources.

NUCLEAR ENERGY – INDIAN AND WORLD SCENERIO

Nuclear energy is a man-made one developed from naturally occurring atomic minerals. Man has a good control over this type of energy. The nuclear chain reactions can be conducted in a reactor to a sufficient degree of accuracy and exactness. The energy is unlimited and sustainable ; and it can be harnessed at will. Very economic plants can be erected following latest designs. In addition to extracting the energy, a variety of radio isotopes useful in industry, medicine, agriculture and research can be obtained from the fission products. The safety of the reactors, and safe handling of the atomic materials have to be given great attention. For this purpose, separate Regulatory Organisations are formed and they stipulate safety precautions. The only disadvantage while securing the nuclear power is the accumulation of radioactive waste. This may cause some environmental problems that are to be tackled with care.

The World output of nuclear electricity is reckoned at 2400 billion kWh/y. This comes from 430 nuclear power reactors situated in 31 countries. The Indian share of nuclear power development is to a mere 1.3 per cent of world output. The country produces a total of 2720 MW electricity from its 14 reactors. The first atomic power plant was erected at Tarapur in Maharashtra state with the assistance from Canada. Since then as many as 10 nuclear power stations were built in India. This power is linked up to the regular grid maintained by the Central Electric Authority of India. In Table 12.2 are recorded details of nuclear power stations, their location, capacity of production.

Table 12.2 Nuclear power stations in India, 2001

Power Station	Location	State	Capacity, MW
Kaiga	Kaiga	Karnataka	440
Madras	Kalpakam	Tamilnadu	470
Narora	Bulandshar dt.	Uttar Pradesh	470
Kakrapar	Surat dt.	Gujarat	440
Rajasthan	Kota	Rajasthan	880

India's nuclear power development programme was entrusted to the Atomic Energy Commission. The Commission deals with all aspects on atomic mineral processing, usage in the reactors, power generation in various plants, isotopes development and their end uses, installation of new plants, research and development activities, training and sponsoring various technical programmes. The safety and regulatory affairs are looked after by a separate body known as Atomic Regulatory Body. Under the Atomic Energy commission a chain of about 50 establishments are functioning at various places in India.

Questions

1. Narrate the various scientific events that led to the discovery of Einstein's equation $E = mc^2$, an expression that explains the quantum of nuclear energy released during fission reactions.

2. Give an account of the 'fissionable Uranium', its mining, extraction and purification and fabrication of fuel rods.

3. Draw a line sketch of a Pressurised water reactor that is employed in nuclear electricity generation.

4. How does a Fast breeder reactor function?

5. Explain Nuclear Fusion reactions that occur between light elements and discuss the advantages of fusion energy over the fission energy.

6. What are the ways of disposal of nuclear waste fuels?

7. Give a brief account of the Nuclear reactor disasters that took place throughout the world, with specific comments on Chernobyl-4 Ukraine reactor disaster of 1986 or Khusk nuclear submarine disaster, 2000.

8. Write commentary on World Nuclear Power programme and its prospects in future.

13

CONVERSION OF MATTER INTO MORE USEFUL FORMS

Broadly speaking the energy resources that are in the form of matter are few as compared to the natural global setting of the environment. All fossil fuels, namely coal, petroleum and natural gas come under the category of matter. The biomass is matter, so also the uranium. A typical feature of the matter is that it is constituted of molecules and has physical state e.g., solid, liquid or gas. The matter also has a shape and exert pressure. It has intrinsic energy in its bonds, hence it is energy delivering in nature. However, the quantum of this energy may sometimes be lower than the energy generated by its 'transformed' or 'modified' products. One should look for those types of modified matter that possess higher energy potential and easy workability than the original matter from which it has generated. In the following paragraphs a few of the 'superior' energy-rendering materials derived from 'less' energy-rendering matter have been described.

WATER GAS

Some varieties of coal show very high ash content and thus its calorific value is considerably low. Such types of coals can be converted to water gas according to the methods specified under Chapte 4. Coal. The resultant water gas has higher calorific value on account of its hydrogen content in the composition and it can be blended with gaseous hydrocarbons to afford 'carburetted water gas'.

Producer Gas: Inferior types of coal such as peat or carbonaceous tars when subjected to processing in a producer generate the high calorific producer gas. Both water gas and producer gas have been considered as cleaner fuels as compared to the inferior coals from which they have derived.

SYNTHETIC GAS

The raw materials for this category of synthetic oils are the tar sands and oil shale. The recovery of petroleum from these types of bituminous oils is difficult and hence they are subjected to thermal cracking when they yield synthetic oil. Synthetic oil is easy to handle and has many fuel applications.

It is noteworthy in this connection that even some petroleum fractions when cracked afford synthetic fuels with high octane number.

BIOGAS/METHANE

The starting material for the production of biogas/methane is the biomass. When treated in digesters with methanogenic bacteria, it yields this gas. Methane has high calorific value (45MJ/Kg) compared to biomass (9MJ/Kg) and is an energy product.

METHANOL AND ETHANOL

Both these substances are obtained from biomass. The distillation of wood chips produce methanol directly. Methanol also is produced indirectly through the oxidation of methane. The reaction between carbon monoxide and hydrogen (both products of coal degradation) at high pressures and moderate temperatures in an autoclave results in the formation of methanol.

Ethanol is obtained in the fermentation of many foodstuffs as well as food crops. Certain yeasts render the fermentation process to proceed up to the stage of ethanol formation. Both the substances have high calorific values and they can also be blended with gasoline. Hence they are branded as bio-fuels.

Synthetic Gas from Municipal Garbage: Treatment of municipal garbage with specific bacteria afford a synthetic gas that can be used for generating steam. The steam is made to move the turbine for producing electricity.

Fischer-Tropsch Synthesis of Paraffin Hydrocarbons: Straight chain paraffin hydrocarbons constitute several blends of petroleum fuels as they improve the octane number. It is possible to synthesise these chemicals on a large scale employing the reaction between hydrogen and carbon monoxide under the controlled conditions laid down in Fischer-Tropsch synthesis. Some of the modifications in the synthesis also led to the discovery of "synthetic gasoline" during the second world war time.

HYDROGEN

This gas with high profile as a fuel is the end product in the bio-transformation of biomass. It has tremendous applications both as a fuel and in synthesis.

In Table 13.1 are shown the comparative heat values of the above-mentioned fuel gases produced through various processes from the raw resources.

Table 13.1 *Comparative heat values of some fuels*

Fuel	Heat value (MJ/Kg)
Petrol	51-52
Methane	45
Propane/butane	50
Methanol	24
Hydrogen	142
Water gas/producer gas	~ 300 Btu/cu.ft

Source: *S. Rao and B.P. Parulekar, 1997.*

Cleavage of Water using Waste Heat : Though not a matter that could be transformed into useful products, waste heat is not let off from the nuclear reactors. It is utilised for cleaving the water molecule at high temperatures above 1700 °C to obtain hydrogen, the universal fuel.

Questions

1. Explain how the raw natural resources can be converted into more-utility high calorific products through chemical transformations.

2. How is ethanol manufactured? What is 'gashol' ?

3. Write notes on the utility of hydrogen as a clean fuel.

14

STORAGE OF ENERGY

The energy produced from various sources of Nature is utilised by the mankind in many ways. After doing so, some excessive energy may be left unused. This will otherwise go waste into the Nature. In order to hold the energy for any future use it is necessary to store it in some form. At the time of need we can extract the same and perform the work. Many natural resources such as fossil fuels and biomass are aleready termed as the 'reserved'/'stored' forms of solar energy. The foods which living beings consume are also 'stored' energy products. Synthetically fabricated electrolytic cells are stored form of electrical energy.

Ultimately the most abundant thermal energy and the electrical power are the ones widely stored in practice. The generation of heat is through the use of fossil fuels or nuclear energy. If one cannot utilise all the heat produced in a system, the burning of the fuel is restricted. Heat arising from solar energy however can be preserved by exchanging it with a fluid that stores it without changing its physical state.

The pre-conditions for its storage are:

- The storage temperature range has to be decided much earlier.

- The storage system should have the capacity to hold the energy intact, as otherwise it will hamper the other mechanisms in the functioning.

- Heat losses from the storage system has to be maintained at lowest order.

- The heat storage system should be cost-effective.

There are three mechanisms through which we can preserve the heat in the systems. These are:

1. Direct heating of a solid or liquid medium that do not change the phase within the specified range. This is termed as 'sensible heat storage'.

2. Direct heating of a material that undergoes a phase change usually from solid to liquid. In this process, the heat is stored as 'latent heat' and is recoverable when the material reverts back from liquid to solid.

3. A chemical reaction occurs under the influence of heat resulting in the products. When the products react again the same heat is released back.

Some of these principles have been applied in designing a heat storage system.

The use of energy may not be continuous. In the automobile transportation sector, for example, more energy is consumed by the vehicles during the peak hours of morning and evening. During the afternoon the number of automobiles operating on road would considerably come down. As a result, low quantity of fuel would be consumed, hence low energy use. Similarly in the electric power sector, the industry takes up large quantity of power during the day time as compared to the night. Eventhough excessive electricity is generated in the power plants, the consumption rate is lower during nights and most of the power remains for use.

From these considerations, it is imperative that proper spacing of energy utilisation has to be made and excess energy has to be stored appropriately. For the managers of energy producers, storage of energy is an equally challenging task as that of production. In this chapter a few established techniques of storage of energy have been described.

STORAGE OF HYDRO ENERGY

Pumped Storage

Excessive hydro energy generated in the power plants is properly stored adapting the 'pumped storage' technique. Usually the dams across rivers generate significant quantity of electric power. The capacity of electricity production is very high during the peak rainy season. Part of this is utilised for industrial , agricultural and domestic purposes. The excess energy is used to pump the water back into a reservoir constructed at an elevated location nearby. This enables the water to be kept at a great height that renders potential energy for the fluid. In a way, the electric power is stored in water as potential energy that can be drawn again at will. It has been established that the recovery of energy from 'pumped storage' is almost cent per cent even taking into consideration the evaporation of water.

A schematic diagram of pumped storage facility is shown in Fig. 14.1 and such a facility was created at the Nagarjuna sagar dam.

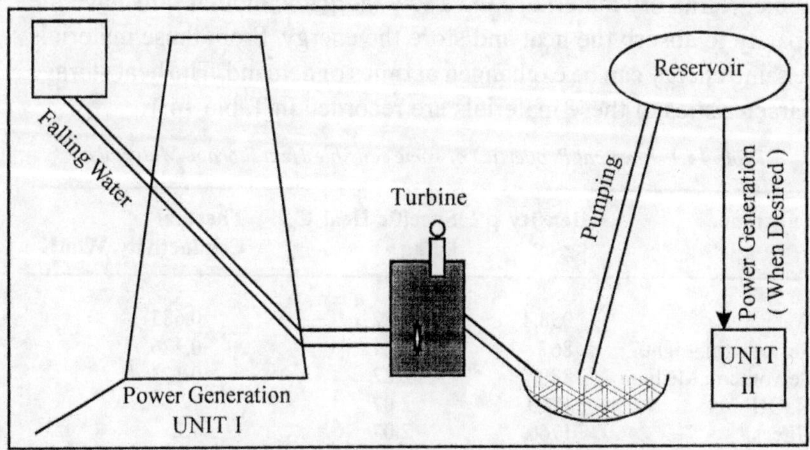

Fig. 14.1 Pumped Storage - Power Generation

Storage Batteries

Large storage batteries are arranged in an array at the electric power plants to supply additional power during periods of demand. The batteries get discharged after they serve the purpose. During the night time, when the demand for electric power is less, these are recharged with the excess hydro power that comes from the hydro power station. The technical details of charging- recharging of secondary cells have been discussed in chapter 11.

Storage of Solar Energy

This is the most abundant energy available on earth and is delivered by Sun during day time alone. The energy has to be utilised as much as we can when it is incident on the earth; however, if we store the same adapting some measures, we will be in a position to draw it during night time too. The following are the systems.

Heat Storage Systems

The incident solar radiation on several liquids or solids cause thermal changes by increasing the temperature without changing its phase. The materials that act as heat preservation bodies are thermal storage systems and the phenomenon is called 'sensible heating'. Water, synthetic heat-transformer oils e.g., Dowtherm, Therminol, Servotherm etc., and rock, pebble, refractory materials, and a few inorganic molten salts have the capacity to absorb the heat and store the energy. From these materials the same energy can be exchanged at times of demand. The heat storage characteristics of these materials are recorded in Table 14.1.

Table 14.1 *Physical Properties of some sensible Heat Storage Materials*

Substance	Density ρ kg/m^3	Specific Heat C_p kJ/kg	Thermal Conductivity, W/mK
Water+	958.4	4.22	0.683
Servotherm, Light	867	2.24	0.126
Servotherm,Medium	880	2.22	0.123
NaOH*	1780	2.07	1.2
Hiteck#	1760	2.07	1.2
Rocks	2245	0.71-0.92	0.13

Contd...

Substance	Density ρ kg/m^3	Specific Heat C_p kJ/kg	Thermal Conductivity, W/mK
Pebbles	1350	0.90	0.85
Magnesium Oxide	3575	1.06	10.5
Aluminium Oxide	4000	1.02	6.3
Silicon Oxide	2600	1.26	2.3

Source: SP. Sukhatme, 1988

+   Properties at 100°C

*   Properties at 320°C, Molten State

\#   Properties at 400°C, Hitec is 40% $NaNO_2$, 7% $NaNO_3$, 53%KNO_2
     by weight

The solar radiation produces heat on the flat plate collector which is exchanged with water. The water attains a temperature upto 100 °C usually and thereafter this is diverted to hot water storage tank. These tanks are built inside a building or in the yard. The tank has to be insulated upto 20 cm thickness with glass wool. The size of the tank is dependent on the area of collection of the solar radiation. Generally for one square metre collection area of solar radiation the tank capacity can be as large as 100 litres.

The temperature of the stored hot water can be maintained slightly above 100 °C when the storage tanks are pressurised. This will enable drawing more energy from the hot water at times of necessity. A hot water storage tank of 5000 litres capacity was constructed at the Indian Institute of Technology, Mumbai .

When high boiling mineral oils such as Dowtherm or Therminol or Servotherm are used for extracting the heat from the solar radiation collector instead of water, temperatures ranging from 100-300 °C can be preserved in these heat transfer fluids. Also, a few molten inorganic eutectic mixtures such as HITEC with $NaNO_3$-KNO_3 (47:53 by weight) can retain the heat within the fluid above 300 °C.

Natural rocks and pebbles have the capacity to store the heat collected from solar sources. The devices are the usual collectors in conjunction with solar air heaters. A line sketch of one system is shown in Fig. 14.2.

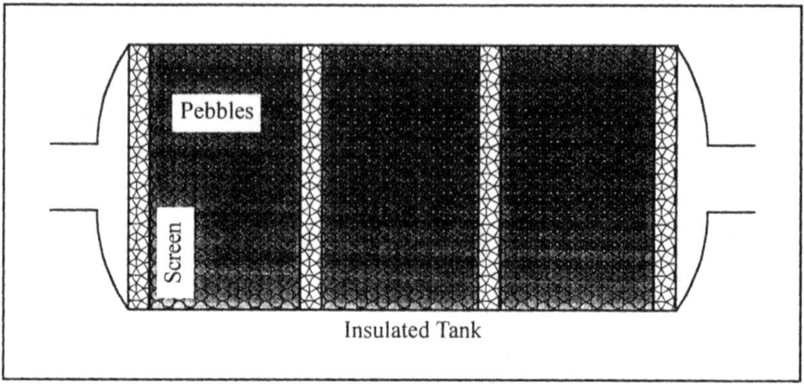

Fig. 14.2 *Heat Storage on Rocks*

The rocks and pebbles store heat beyond 100 $^{\circ}$C and sometimes upto even 1000 $^{\circ}$C. The physical properties of some heat transfer liquids, minerals and rocks are recorded in Table 14.2. Natural minerals such as magnesia, alumina and silica also possess heat storage capacity.

The amount of heat stored in the above mentioned media is dependent on the temperature change of the material from the original condition to the hot condition and this can be represented as T2 - T1. This is called the 'temperature swing' and if this figure is very high the energy that is intrinsically available will be considerable. Many systems have been developed utilising this phenomenon and were put to use. Some pressurised hot water tanks are potential sources of stored energy.

Latent Heat Storage

Certain materials inherently have the capacity to absorb the heat (from solar radiation) in the molten state and give out the same when the material freezes. The mechanism of storage of energy is different from the "sensible heating" system in the sense that all the molecules in molten state equally contribute to the storage. When the material freezes the outer layer solidifies first thereby heat transfer between the storage material and the working fluid, usually a gas, is made over the surface first and later with the core matrix of the stored material. However, the latent heat storage by heat exchange of the latent heat of fusion is considered a compact one.

The following substances display the latent heat storage mechanism.

Table 14.2 Materials displaying Latent Heat Storage of Energy

Substance	mp^{o}C	Heat of Fusion, kJ/kg
water, H_2O	0	335
Na_2SO .$10H_2O$ + NaCl/KCl	13-18	—
$CaCl_2$.$6H_2O$	30	168
Na_2SO .$10H_2O$	32	241
Paraffin wax	46	209
$Ba(OH)_2$.$8H_2O$	82	266
Naphthalene	83	148
$Mg(NO_3)_2$.$6H_2O$	89	167
$NaNO_3$	310	173
NaOH	320	159
LiF-NaF-MgF(46-44-10)	632	814
$NaF-ZnF_2$ (63-37)	640	599
$NaF-MgF_2$ (75-25)	832	626
LiF	845	1044

Source: SP. Sukhatme, 1988.

The fundamental requisites for a material to be capable of causing latent heat storage are:

1. A convenient melting point for the substance in the prospective working range

2. High latent heat of fusion in order to pass on the heat energy to the working fluid

3. Small volume change during transformation of phases

4. Least supercooling or superheating characteristics in the two phase changes

5. High thermal conductivity in the two phases of occurrence

6. Low vapour pressure at the working temperatures

7. Non-corrosive nature.

Amongst the materials considered above, those having melting points above 300 ^{o}C are worth selecting for energy conversion uses and the ones in the melting point range 30-89°C for space heating applications.

Different stacking arrangements of the latent heat storage materials provide efficient heat exchange capability for a gaseous working fluid.

Thermochemical Storage

A sequence of one forward and reverse reactions could store the energy processed from solar radiation and give out the same when required. It is as follows:

Forward Reaction

$$A + B \rightarrow (Reactor) \rightarrow X + Y \rightarrow Storage\ Tank$$

(heat collection from collector, endothermic)

Reverse Reaction

$$Storage\ tank \rightarrow X + Y \rightarrow Reactor \rightarrow A + B$$

(heat release for application, exothermic)

A few suitable thermochemical reactions for heat storage are shown below:

Reaction	Forward reaction C	Reverse reaction C	Energy Stored per unit volume of Storage Material, kJ/cu.meter
$CH_4 + H_2O \rightarrow CO + 3H_2$	780	610	209.4×10
$SO_3 \Leftrightarrow SO_2 + \frac{1}{2}O_2$	1025	590	460.6×10
$NH_4HSO_3 \Leftrightarrow NH_3 + H_2O + SO_3$	498	435	2143.7×10

Source: SP.Sukhatme, 1988.

Some of the essential requirements in choosing the reactions are:

1. Both forward and reverse reactions are governed by the capacity of heat obtainable from the solar collector and the temperature of these reactions are dependent on the specific type of solar radiation collection.

2. The efficiency of the collector would be high when the temperatures of forward and reverse reactions are close.

3. If the products X and Y happen to be liquids, and possess high values of energy absorption/unit volume, then the storage space can be minimised.

4. Both forward and reverse reactions shall preferably be fast, with the possibility of no side reactions.

These reactions are successfully employed in the solar power plants and heating systems.

Photovoltaic Conversion-Solar Cells

The device that converts solar radiation into direct current is a 'solar cell'. It is made up of thin wafers of n-type silicon (0.2μ) and p-type silicon (300μ) thickness and diameters of 3-6 cm. The electrodes are made up of titanium- silver and are soldered on both front as well as back sides of the block. The solar cell has the capability of converting the sunlight directly into electricity (DC) when it is incident on the surface. The solar cell works on bright and diffused light as well and it does not have any moving parts to generate the current. It remains static in the 'arranged array' which is positioned in an angle to receive maximum solar radiation. When light falls on the cell it absorbs the same and produces pairs of +ve and -ve charges. These are subsequently separated at the p-n junction resulting in the development of electricity which is passed on through the two electrodes outside. The diagram of a solar PV cell is already shown in Fig. 5.6.

Limitations

An important consideration for the choice of the semiconductor material to be used in the photovoltaic cells is the inherent band gap energy (E_g). A band gap energy which is considered reasonable for achieving maximum efficiency of current generation is possessed by the silicon wafers and this value is 1.20 ev. Similar ev is shown by electronic materials such as gallium arsenide, cadmium telluride and cadmium sulphide also.

The sunlight possesses energy E emitted as photons. If the E value of sunlight happens to be less than the band gap energy Eg of the semiconductor material, the photons are not effectively absorbed, whereas if the photons with energy greater than the Eg of the material, that is E >Eg , the partial energy E - Eg is spent as heat.

An inter-relationship between the current- voltage offered by a solar cell determines the efficiency of the cell. If one draws a graph between these two parameters, the value of current I* obtained at zero voltage is termed the 'short circuit current' whereas the voltage V* arising in an open circuit is termed 'open circuit voltage'. The product of I* and V* is the ideal power of the cell but the large available power is represented by the area of the most possibly accomodated rectangle within the voltage- current curve. For the largest rectangle fitted within the curve, if V# and I# are the maximum voltage and current respectively, the most available power is denoted by V#.I#. The ratio of the maximum available power to the ideal power is the 'fill factor'.

$$\text{Fill factor (F)} = V\#.I\# / V*.I*$$

A satisfactory fill factor between 0.65- 0.80 is shown by silicon solar cells with V* 450-600 mv and I* 30-50 mA/ cm2.

Photovoltaic Array

When several solar cells are mounted on a panel supported in such a fashion that maximum solar radiation is incident on the cells that system is called a 'photovoltaic array'. The individual solar cells in an array comprise of anything between 24 to 100. Such a system has the capability of providing light or pumping water from a reservoir and similar tasks. A line sketch of a solar cell array is shown in Fig. 14.3. An advancement of this system is the concept- design of 'satellite solar power station'. The satellite solar power station comprising of innumerable cells has to be constructed on a satellite which is placed in a 'geosynchronous' orbit far above the earth. The cells in the array occupy several square kilometers area and therefore the station is expected to deliver large quantity of electric power.

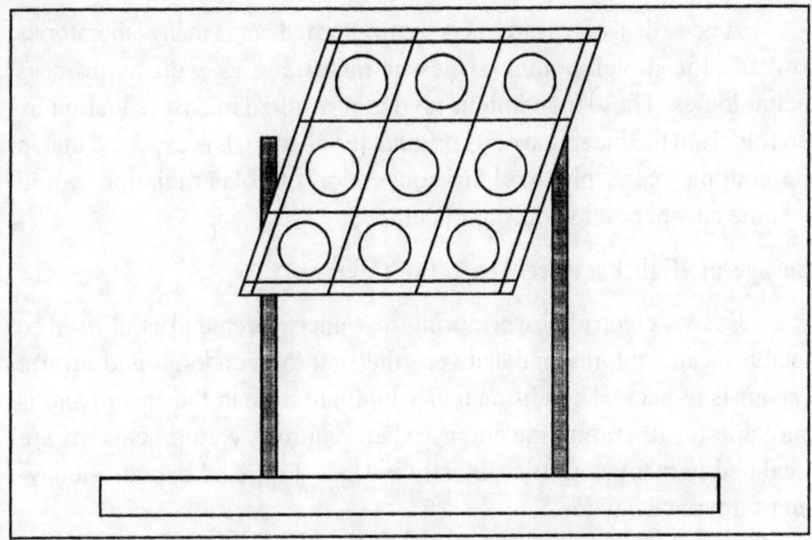

Fig. 14.3 Line Diagram of Solar PV Array

The DC supply that is generated upon the incidence of Sun's rays can be fed to microwave oscillators that in turn is transferred to the antenas on the earth. Then the microwave energy is to be converted into dc supply on the earth that can be used for lighting, domestic and industrial uses.

It is thus clear that the solar energy is eternally stored in the semiconductor cells in the form of electricity. Also, it can be drawn whenever any requirement arises. Since the cells or arrays are small in size, they can be fabricated and lifted to difficult areas. These are therefore installed in mountain areas where it is difficult to lay the cables and supply the normal electricity. Many military uses are found for the photovoltaic cells in camps, field areas and control chambers. The spacecrafts namely Mariner and Skylab have the facilities for photovoltaic conversion of solar energy.

The cost of photovoltaic development of dc supply is very high. This is chiefly due to the processing of silicon wafers wherein automation for continuous supply is difficult. In order to obtain good quality monocrystalline piece a lot of effort has to be put and nearly three fourths of the material goes waste, leaving behind only very little for the process.

A good deal of research is presently carried out in many laboratories towards the development of newer materials, as well as process technologies. These innovations no doubt resulted in cost reduction by ten fold from the inception of its original production. It is expected that in the coming years, photovoltaic conversion of solar radiation would become cheaper and widely applicable.

Storage of High Pressure Air in the Caverns

Excess electric power during low energy demand is utilised to operate an air compressor that diverts the air into an underground cavern. The air is in packed condition under high pressure in the cavern and is available for liberating the energy when required. Natural caverns are escalated in many countries, mostly in USA. Fig. 14.4 depicts the lay-out of a typical cavern.

Fig. 14.4 *Storage of Hot air in Caverns*

The HP air is slowly released from the cavern, heated by gas and then subjected to drive the turbine. Electricity is generated at this end, which is used as per the requirement. The McIntosh, Albama turbine plant runs on the heated air coming out of a cavern.

Shaft-Driven Fly Wheel

Fly wheels with connective shaft drive are commonly found in steam engines and locomotives. A fly wheel is a heavy metal wheel in which most of its weight is concentrated in the circumference. The wheel is attached by means of a drive shaft to receive the energy of a moving piston. Now the device works as a store of electric power. A revolving wheel stores the mechanical energy by using its own mass in the circumference to propel and accelerate when put in a little motion. Thus if a fly wheel swings into action at rapid rotation, it performs work similar to a turbine to generate electric power. The rotation of a speeding fly wheel is thus the stored electric power in locomotives. In Fig. 14.5 is shown a sketch of a fly wheel power shaft.

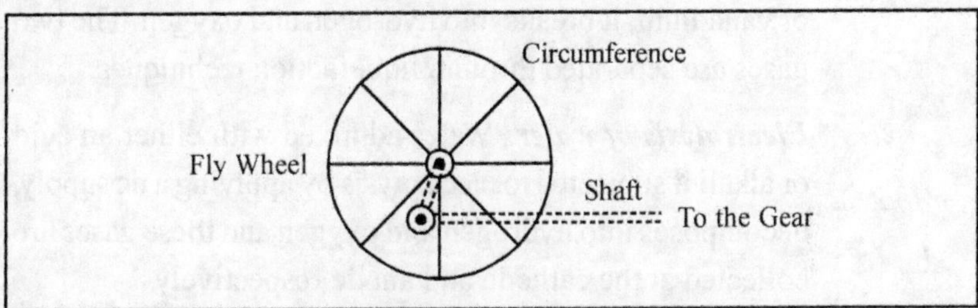

Fig. 14.5 Line Sketch of a Fly Wheel

Chemical Conversion Method: Certain energy producing substances such as oxygen or hydrogen can be chemically converted into compact forms in order to store large quantities of energy within them. These substances are the 'peroxides' and 'hydrides', hydrazine, organic azo compounds etc. Decomposition of these substances release enormous quantities of energy that can be used in several metallurgical processes, propulsion of missiles and navigation of torpedoes etc. Using appropriate methods, the energy can be extracted in different ways from these stocks. The preparation and properties of a few materials are discussed hereunder.

HYDROGEN

The chemical element 'hydrogen' is an embodiment of energy in modern times. It is the first element in the periodic table, and the most abundant one on earth. Henry Cavendish discovered this light gas in the year 1766. It has an atomic number 1 and atomic weight 1.008 and density 0.05459. The gas is inflammable and with oxygen it turns into water. Besides, hydrogen is the constituent of millions of organic compounds available in nature.

Production

The following are the general methods of large- scale production of hydrogen.

1. ***Thermal dissociation of water :*** When water is subjected to heat at high temperatures e.g., 2000 ^{o}C either alone or in presence of catalysts such as oxides of iron , molybdenum or vanadium, it breaks into hydrogen and oxygen. The two gases are separated through liquefaction techniques.

2. ***Electrolysis of water:*** Water admixed with either an acid or alkali if subjected to electrolysis by applying a dc supply, decomposes into hydrogen and oxygen and these gases are collected at the cathode and anode respectively.

3. ***Thermal reforming of coal:*** If coal is heated over passing superheated steam, it generates a mixture of CO and H_2 and this is known as 'water gas'. Hydrogen is separated from the mixture suitably.

4. ***Conversion of biomass:*** It is well-known that biomass produces methane in a number of ways. Superheated steam if reacted with methane produces hydrogen as one of the products of conversion. In this way large quantities of hydrogen is produced. Interestingly this is the cheapest of all the methods.

The hydrogen produced by several of the abovementioned techniques corresponds to approximately 2% of the world's energy consumption. Some of the inputs essential for the production are:

- A source of heat for dissociating water molecules. Solar reflectors that concentrate the heat at one point to a temperature of about 2000 °C placed in a desert area would ensure constant supply of heat for the process; alternatively, waste heat recovered from a nuclear power reactor also supplies significant heat. The use of metal oxide catalysts would lower the reaction rate to approximately 700-800 °C.

- The source for generation of dc electric supply can be usually from one of these natural resources, eg. solar power concentrated by reflectors in African deserts, solar photovoltaic electricity; hydropower (Quebec, Canada); wind power in Scotish Islands; geothermal energy in Iceland etc. A different approach shall be to adapt the latest versatile Solid Polymer Electrolysis (SPE) for electrolysing the water into hydrogen and oxygen, both on a large scale effectively (details shown separately below).

- Descriptions of thermal reforming of coal into water gas and the conversion of biomass through methane to hydrogen are given in this text.

SOLID POLYMER ELECTROLYSIS

Solid Polymer Electrolysis (SPE) is a new version of electrolysis of water to afford hydrogen and oxygen. Developed in 1979 by General Electric Co for NASA this technique found many applications in nuclear submarines and manned spacecrafts. Today this technique finds use in more than 12 applications where either pure hydrogen or oxygen is in demand. SPE cells function for over 150 million cell-hours with an assured 40 million hours of 'system' operation.

The cells operate under stringent conditions of atmosphere, especially at 3000 psi pressures in the cabins and have great endurance.

Conventional electrolysers use aqueous solutions of either an acid or alkali as electrolyte, besides the anode and cathode. The electrolyte is broken into ions which polarise to the electrodes causing separation of hydrogen and oxygen. The electrolyte needs replacement as the process progresses for some hours; also, the concentration of the electrolyte has to be checked occasionally. They are also caustic substances and are to be handled carefully. Spillage of these substances in the cabin spaces is unwelcome as the materials are not only caustic but are also corrosive; hence the body material is damaged.

In the case of Solid Polymer Electrolysis no liquid electrolyte is employed. This obviates the risk of spillage, leakage and handling problems. The electrolyte is a solid fabricated from a 'conducting polymer'. It is made into a compact solid mass by cladding films of noble metal between the film sections of the polymer. One end of the metal serve as an anode while the other the cathode. Conducting polymers of the type, poly tetrafluoroethylene sulphonic acid or poly pyrrole or poly aniline are commonly used in thin film forms. They are prepared from the monomers through several steps of polymerisation. In the solid form these materials are quite rigid and possess electrical conductivity that is a requisite in the process. A variety of conducting polymers have been nowadays available on commercial scale and these find many applications in the fabrication of electrolytic cells and sensors.

A diagram of a SPE cell is shown in Figure 14.6. The cell is compact and rugged constructed with conducting polymer membranes and noble metal electrodes. The inter membrane distance of polymer and metal cladding is so little that it is possible to apply high current density and achieve maximum efficiency. This is a positive feature of solid polymer electrolyser beside its simplicity. The membrane of polymer also serves as a good separator for the evolved gases. The cell has long endurance.

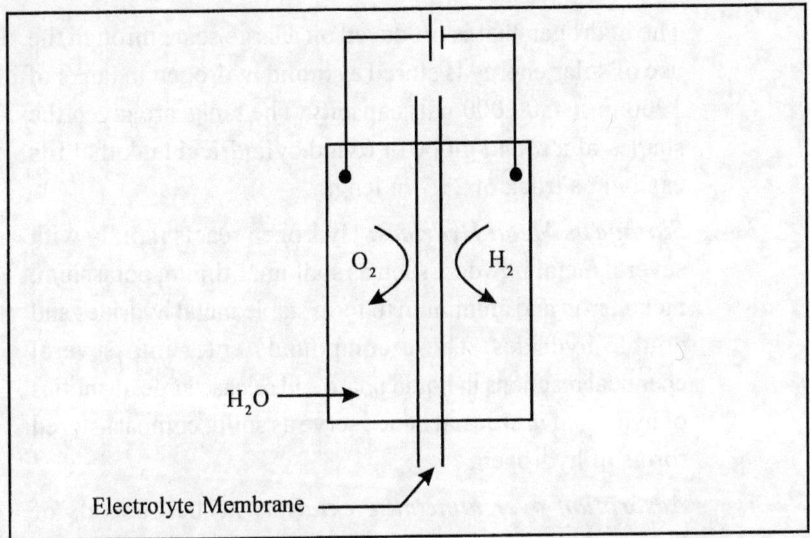

Fig. 14.6 *Line Diagram of Solid Polymer Electrolysis (SPE) Cell*

Photovoltaic (Solar) Cells

The device which delivers a dc supply to be used for electrolysing water is the 'photovoltaic or solar cell'. This has the capability of storing the sun's radiation and giving the output in the form of electric current at any needy time. A detailed review of the photovoltaics is made in the earlier pages of this chapter.

STORAGE OF HYDROGEN

Hydrogen is stored in three ways. These are:

1. ***Storage in Pressure Vessels:*** The gas that is produced either in the steam- reforming of methane or during electrolysis is led into a chamber where it is compressed and filled into pressure vessels. These are steel gas cylinders that hold up hydrogen up to $7.45m^3$($263\ ft^3$) at pressure of 164 atmospheres. The purity of the gas is 99.99%. If ultra-high purity hydrogen is required an additional step of passing the gas through a molecular sieve to hold back large molecular gases is introduced before the compression of the bulk gas.

2. The hydrogen that is produced on a large scale through the use of solar energy is stored as liquid hydrogen in tanks of 1900 m^3 (~500,000 gal) capacity. The tanks are given the shapes of a round globe or round cylindrical tank that fits easily in a truck of 23.7 m length.

3. ***Storage as Metal Hydrides:*** Hydrogen reacts rapidly with several metal powders such as sodium, lithium, potassium, nickel, iron and aluminum to form stable metal hydrides and double hydrides. These compounds enter into several chemical reactions in liquid phase and release large quantities of hydrogen *in situ* and hence serve as solid, compact stored forms of hydrogen.

4. ***Adsorption over bimetallic catalysts :*** The bimetals of the type Iron-Titanium, Iron-Lanthanum, Iron-Magnesium, Iron-Nickel have the capacity to take up hydrogen (with the generation of heat) at normal temperature and pressure forming grannular particles of 475μm (4 mesh) dimension. The process of adsorption is invariably exothermic; the bimetals could take up from 4 to 100 times the quantity of hydrogen within its structure and the resulting solid form is stable and compact. It is a potential reserve of hydrogen in the solid state.

    The reaction can be represented as:

    $$FeTi + H_2 \rightarrow FeTiH_2$$

5. The absorption of hydrogen can be achieved satisfactorily at room temperature and gas pressure of 34 atmospheres in general. The iron-titanium alloy has to be pre-heated to 300°C before it is charged into the autoclave. An increase in volume of the alloy is noticed during the reaction together with embrittlement. A Fe-Ti-Mn alloy with manganese content of 10-15% is an improved bimetal that is reactive with hydrogen and can be recycled many times.

The binary/ternary hydrides thus formed are usually enclosed in steel cylindrical containers to hold about 560 kg of the material and transported. This type of storage on bimetals is considered superior to the liquefaction or the pressure vessel storage. Generally the hydrogen content in the solid mass would be about 5-9% by weight of the bimetal. The release of hydrogen from $FeTiH_2$ can be achieved at 770 mg of Hg at 30-80 $^{\circ}$C. The purity of the released gas is very good, however care has to be taken in not admitting any fine solid particulates during the collection of the gas.

Hydrogen as Fuel

Hydrogen burns in presence of oxygen liberating heat equivalent to 29,000 cal/cc(2,050 Joules/gm). The product of combustion is water alone, unlike greenhouse gases in case of all types of fossil fuels. No soot is evolved. The gas can be held in pressure bottles; it can also be liquefied and transported easily. Hydrogen gas can be transported through pipelines to any distances without losses. It can be compacted in high-capacity hydrides or metal alloys which are solids. Lastly hydrogen burners are very clean as the gas does not leave any residues after burning. If blended with natural gas, the formation of carbon dioxide can be minimised during combustion processes. On account of these qualifications the gas can be used as a fuel.

As a fuel, hydrogen enters into two types of combustion; the first direct burning to generate heat that can be used for producing electricity for engines; the second, to participate in a combustion reaction with oxidant in a fuel cell, that ultimately results in dc power supply. The conversion of hydrogen to electricity on burning with an oxidant is very efficient.

Cost

Current cost of production of hydrogen from any of the renewable resources is rather high.

However, with the large scale utilisation of solar energy in desert locations coupled with good supply lines to the nearby industrial areas, the production costs of hydrogen will certainly come down. Widespread usage of hydrogen by nations and tax incentives by the respective governments also bring cheaper hydrogen to people.

FUEL CELLS

A" three step array "consisting of an 'anode' catalyst, polymer electrolyte membrane and a 'cathode' catalyst in a row of an enclosed chamber form the *fuel cell.* When the fuel , hydrogen is fed into the anode side of the cell and the oxidant, oxygen sent to the cathode side, and if both anode and cathode are connected outside by a wire, one finds sufficient amout of electric current generated to light an incandescent bulb. The sequence of migration of protons and electrons of hydrogen in the system is shown in Fig. 14.7.

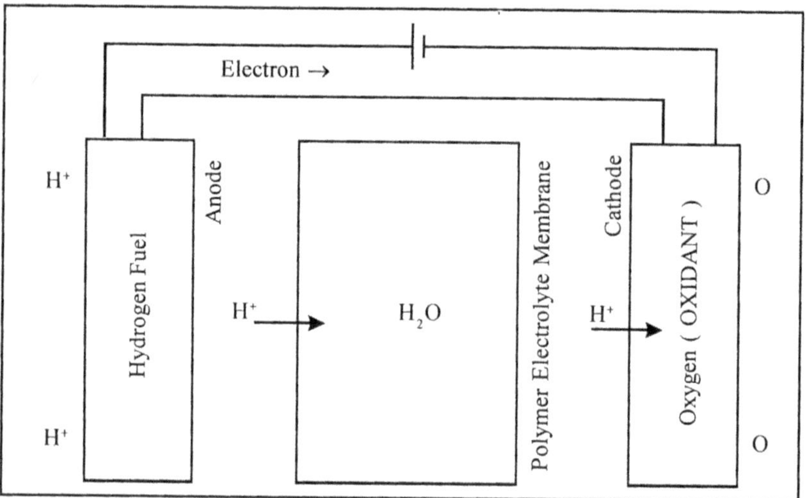

Fig. 14.7 *Line Diagram of Fuel Cell*

Sequence of Migrations

Under the influence of the catalyst at anode, hydrogen atom splits into an electron and proton which take different routes to reach the cathode.

The proton from hydrogen migrates through the electrolyte to the cathode. The electrons create a separate current that reaches the cathode faster and gets ready for reuniting with the H+ and O_2 to form molecular H_2O thus discharging the +ve and the –ve charges. Ultimately this chemical transformation results in the generation of electric power.

The principle of operation of fuel cell is similar to electrolysis of water but in the reverse manner. Under regular electrolysis we cleave water into hydrogen and oxygen on applying a dc supply, whereas in the fuel cell, individual hydrogen and oxygen molecules combine and generate water and dc supply. As we notice, the chemical reaction is ionic, fast and hence the efficiency of consumption of fuel and oxidant is high resulting in efficient power generation. At the same time , the process is pollution-free, as the end product is water alone.

Based on these principles, a variety of fuel cells forming typical electrolytes have been developed since 1952. Many of these found application in automobile, defence equipment and in the electronics for guaranteed small power supply. Some of these systems are described hereunder.

1. *Alkali fuel cells:* A solution of potassium hydroxide in a matrix is the electrolyte and compressed hydrogen and oxygen are the fuel and oxidant respectively in the fuel cell. The OH ions from the cathode side migrate towards the anode first and react with hydrogen at the anode releasing the electrons. The electrons thus released at the anode travel through the external circuit to the cathode and react with oxygen and water to produce more OH ions in the electrolyte. The electrolyte functions effectively at temperatures between 150-200 $^{\circ}$C.

   The electric output from these cells is in the range of 300 W to 5 kW with an overall efficiency of 70%. No pollution takes place and pure water is the end product. The cells were used by NASA on spacecraft Apollo for providing electricity as well as drinking water.

Besides, these fuel cells found use in transport sector for moving vehicles such as taxis, delivery vans, airport tugs, power boats etc. The Union Carbide Co. designed a fuel cell - powered motor cycle in 1967 and demonstrated the exhibit. Several defence equipment such as radar, undersea machines utilise the power from fuel cell and function effectively. A British company by name Zero Emission Vehicle Company (ZEVCO) introduced in London, the first proto-type electric taxi powered on a 5000W fuel cell. It is conceived that these cells find application in the movement of 'stealth' vehicles operated by the armed forces.

2. ***Phosphoric Acid Fuel Cells:*** Selection of phosphoric acid electrolyte in the fuel cell dates back to 1961 when an 'intermediate' fuel cell was reported by Elmore and Tanner and claimed as running on air oxidant rather than oxygen very efffciently. Later improvements brought out a sturdy model that is capable of even running an electric bus.

In the Phosphoric Acid Fuel Cell (PAFC) the electrodes are made up of platinum to conduct the current and act as catalyst. The electrolyte, phosphoric acid, 35 parts mixed with silica powder,65 parts, is the matrix. The hydrogen proton which generates at the anode migrates through the electrolyte to the cathode; the electron passes through the external circuit as current reaches the cathode and combines with the hydrogen proton and oxygen to form water. These fuel cells are capable of delivering electricity outputs between 200 kW to 1MW at operative temperatures of 150-200 $^{\circ}$C. The cell has very good endurance.

PAFC can function even with the supply of lower grade hydrogen even. The platinum electrodes are not poisoned upto this level of the CO. Sometimes the hydrogen carries a little of carbon monoxide as an impurity and in the fuel cell, the concentration upto 1.5% has no deleterious effect. The overall efficiency of the cell is about 50%.

Today there are more than 200 PAFC systems operating throughout the world. Both stationary and mobile power generators are in use.

A 100 kW PAFC powered electric bus was demonstrated by Toshiba and United Technologies together under a joint project. Even the city of New York has installed many of these systems to augment the regular power supply and as a stand-by at times of emergency black-outs. The fuel cell systems were also introduced in military equipment.

3. ***Molten Carbonate Fuel Cells:*** The researches on the introduction of high temperature electrolytes in the fuel cell development started around 1950s in several countries with the objective of achieving high quality cells. It was hoped that at very high temperatures the interference of gas, carbon monoxide arising from the reforming of hydrocarbons, can be minimised. High temperature solid as well as molten carbonates were simultaneously studied by many scientists. Thus many molten carbonates of lithium, sodium or potassium have shown promising results.

When the carbonates of alkali metals are mixed and heated to above 650°C they melt and form a 'molten carbonate' electrolyte that conducts electricity. The carbonate ions move freely within the liquid from cathode to the anode. The hydrogen at the anode reacts with the carbonate ion (CO_3) producing water, carbon dioxide and electrons. The electrons, as usual, pass through the external circuit and reach the cathode; the carbon dioxide from the anode migrates through the molten liquid towards the cathode. At the cathode, the carbon dioxide and oxygen of air in presence of the electrons combine to liberate the carbonate ion back into the electrolyte besides generating electricity.

The output given by the MCFC systems range from 10 kW to 2 MW electric power. These are mostly stationary structures as the operations involve high temperatures. It was conceived that the systems offer high performance of the order of 60-80% efficiency. The fuels are invariably the hydrocarbons of fossil fuels that offer hydrogen. On account of the high temperatures employed, no poisoning of the catalyst electrode is envisaged; also one can switch over to less-expensive metal electrodes such as nickel for these systems.

On the whole, the fuel cells cost less with the promise of 'high' fuel-to-electricity efficiency. Even the waste heat can be recovered for space heating purposes.

Many power plants were designed by Japanese companies and the Ishikawajima Heavy Industries in Japan operates a 1000 W MCFC. The Fuel Cell Energy Inc. runs 2 MW system at Santa Clara, CA. Miramar Marine Corps Air Station in San Diego operates a 250 kW MCFC since 1997. Besides, there are many utility electricity producing stationary plants running in many automobile industries in USA.

The disadvantage one encounters in these power plants is the corrosion of the machinary. Also, internal chemical reactions between the carbonates result in some loss of carbon dioxide that necessiates the addition (by injection) at the cathode periodically. Some maintenance problems also push up the cost of production of electricity.

4. ***Solid Oxide Fuel Cell :*** This is yet another highly promising type, high power application oriented and useful in stationary electricity generation as well as transport systems. The technology is similar to the MCFC type. It operates at a temperature as high as 1000 °C, so far the highest among fuel cell types. The power generating efficiencies go up to 60-85% at the cell output of 100 kW. The shapes of the fuel cell can be designed in the form of 'array' of small tubes, long tubes or round compressed discs to suit the machinary.

The electrolyte is a solid mixture of zirconium and calcium oxides; this is in the form of solid crystal. Both sides of the solid electrolyte are coated with special porous electrode materials and the cell is shaped. As in any fuel cell, oxygen is fed at cathode; it is converted to the negative ions which travel through the electrolyte to the anode, react with the available positively charged hydrogen, allowing the electron to flow through the external circuit to the cathode and ultimately generating current.

The advantages of the SOFCs are too many; first, they give clean energy with the use of hydrogen obtained by 'reforming' process; second, even air as a source of oxygen can be adapted. Lastly, the waste heat produced in the chemical reactions can be turned into further electric power by operating a steam turbine, or it can be utilised for space heating. The Energy companies in USA and Japan have demonstrated many versions of the SOFC systems for power generation in the range of 25-100 kW.

5. ***Proton Exchange Membrane Fuel Cell :*** These are the only type of fuel cells that operate at near atmospheric temperatures, very compact in construction, sturdy, and possessing high power density. PEM fuel cell technology was originally developed for NASA by the General Electric Company during the periods 1963-1979. The power output from these systems can be varied quickly to meet the user's demand and hence are most suitable for automobiles where a 'quick start-up' is a requirement. The cell outputs are in the range of 50-250 kW.

The electrolyte is a solid polymer membrane that is a conducting type, and capable of permeating the protons of the fuel, hydrogen towards the cathode. The electrodes are made up of noble metals powder, highly dispersed in form and coated on both sides of the solid polymer. The electrolyte is a polymeric material, usually perfluoro ethane sulphonic acid type that is proton exchanging. In the actual process, the hydrogen fed at the anode side splits into a proton and electron. The electron passes through the external circuit to the cathode, while the proton diffuses through the anode side to the cathode side through the electrolyte membrane. At the cathode the electrons participate in the reaction between hydrogen proton and oxygen thus forming water.

PEM fuel cells were first tried on NASA's Gemini project missions for supplying power. Today several automobile manufacturers are evincing keen interest to incorporate the cells in the design of vehicles. Ballard Systems manufactured a 5 kW stationary power system.

The Crane Naval Air Station in Indiana, USA, is using a 250 kW power plant for energy. Many such systems were already in use throughout the world.

6. ***Direct Methanol Fuel Cell:*** This is very similar to the PEM fuel cell in operation except that the fuel hydrogen is given from methanol that is in contact with the anode. The step of 'reforming' to obtain the fuel is thus eliminated in the design. DMFCs also operate at low temperatures, namely 50-120 $^{\circ}$C with an optimum efficiency of 40%. The fuel cell offers small power only and as such it finds use in some electronic equipment such as cellular phones and laptops.

FUTURE

Almost all the major automobile manufacturers in the world today have come up with models of vehicles that function on hybrid power systems. The theme has been welcomed in advanced countries for two reasons; firstly, the fuel of the automobile leaves no pollutant; second, the dependence on fossil fuel has reduced. In fact, many major power generating industries are preferring to switch over to the fuel cell instead of the age-old coal or oil or gas that cause both air and soil pollution. On the domestic front, many utility companies also prefer erecting and distributing the less-cumbersome fuel cell power than the conventional electricity (that comes with many uncertainities). Besides, the applications of fuel cells in ground and air military operations are too many. Even in remote areas these facilities are made to work. The conversion of solar energy through PV and electrolysis to hydrogen enables one to store the gas and put it for use in fuel cells for discharging electric power during night time in unmanned unspecified locations. Walking alongside the manufacturing technology of conducting polymers, the fuel cell technology is expected to grow and offer unlimited power for the world.

Questions

1. How is excessive hydro energy stored? Give an account of the Pumped Storage.

2. Explain the principle involved in Photovoltaic conversion. How are solar cells useful in daily power usage?

3. Write two methods of preparation of hydrogen on industrial scale. Discuss various methods of storage of hydrogen

4. What is Hydrogen Economy?

5. Explain the principle of electricity generation using hydrogen.

15

DIRECT ENERGY CONVERSION

Carnot showed that all engines which convert heat into mechanical work operate by transferring heat from a source at a temperature T_1 to a sink at a lower temperature T_2

and that the efficiency ε of such an engine is given by

$$\varepsilon = \frac{T_1 - T_2}{T_1} \qquad .....(15.1)$$

Since $T_1 - T_2 < T_1$ then ε i.e., the efficiency is less than 100% (of the change in enthalpy of the reaction).

Hence for a given source and sink temperatures, the maximum possible efficiency is prescribed by what is *Carnot limitation* on the conversion of heat into mechanical work. The limitation is *intrinsic*; it cannot be avoided by improvement of the engine design. For e.g., the steam engine working between 356 and 100°C has a maximum efficiency of 41%.

CONVERSION OF CHEMICAL ENERGY INTO ELECTRICAL ENERGY

Apart from the fact that the usual method of producing electricity, the most convenient form in which energy can be made available-is subject to the Carnot limitation, it is clear that it is an indirect method. A chemical reaction produces heat, which causes a gas to expand and thus push forward a part of a machine. This machine then drives another which produces the electricity. It would be preferable to devise a means of converting the energy released during most chemical reactions to electricity without involving a second machine and also preferably without moving parts for these are subject to erosion and mechanical failure. Several such methods called *direct energy conversion methods* are known. In the thermo-ionic converter (Fig.15.1), an electronic conductor is heated till it emits electrons; these eelectrons are taken off by a counter electrode, and a current is made to flow through an external circuit. In the thermoelectric devices (Fig.15.2), two differing materials A and B (metals or semi conductors) are made to form two junctions, each consisting of an A to B contact; one junction is maintained hot and the other cold, and electricity flows between them through a load. In magnetohydrodynamic converters(Fig.15.3), a hot plasma is circulated past the poles of a magnet. A fuel is used to heat the gas and ionize it.

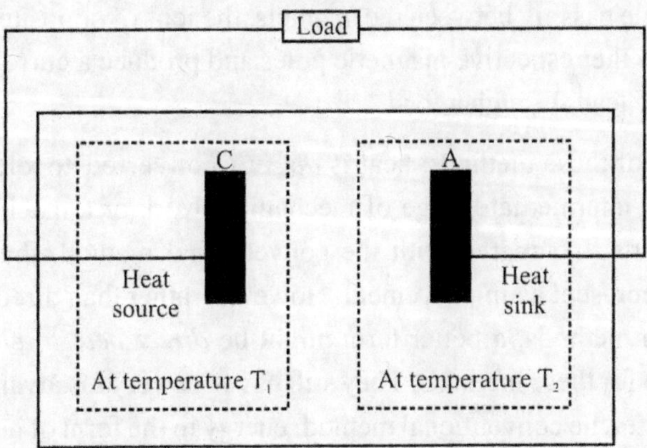

Fig. 15.1 *A schematic representation of a thermoionic converter where C is the emitter cathode and A is the collector anode*

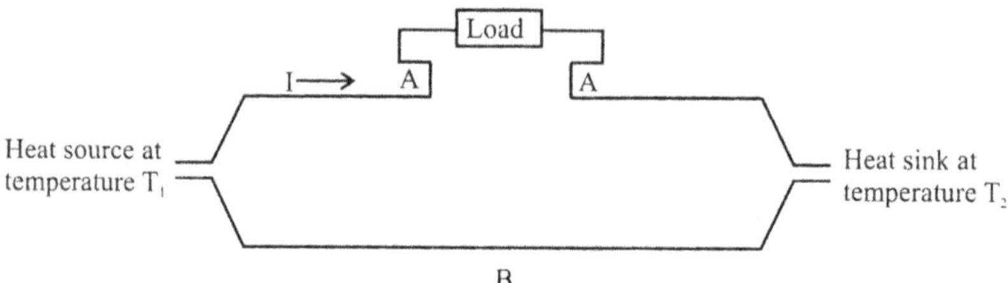

Fig. 15.2 A simple thermoelectric energy converter.

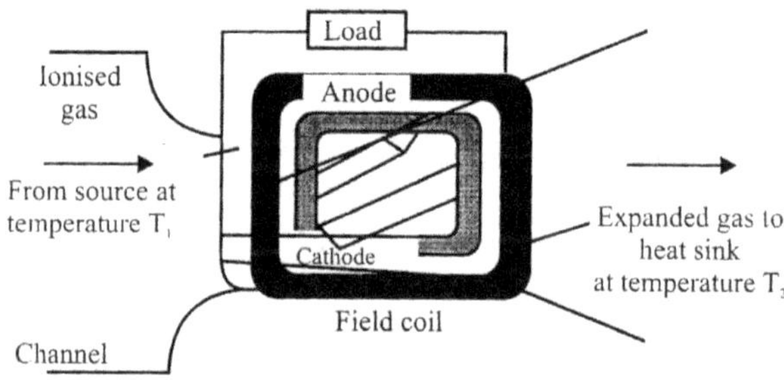

Fig. 15.3 The elements of a magnetohydrodynamic(MHD)engine.

While passing between the magnets, the ions of opposite sign are attracted to the respective magnetic poles and produce a current which can then be lead through a load.

In each of these methods, heat is *directly* converted to electricity without an intermediate stage of mechanical work or a machine with moving parts. Compared with the conventional method, the present methods represent an improvement. However, rather than direct energy conversion methods, a better term might be *direct heat-to-electricity* conversion for these methods. They still have one very disadvantageous property with the conventional method; energy in the form of heat is put in at a high temperature and comes out of the converter at a lower temperature. They are therefore still heat engines subject to the Carnot efficiency limitation; i.e., only a fraction of the energy of the chemical reaction can be turned into electricity, however well the devices are engineered. What is needed is the method in which the energy is released in chemical reactions is indeed converted *directly* to electricity without moving parts, as with the several methods outlined above, but also in a way which *avoids the loss of energy due to the intrinsic carnot limitation.*

DIRECT ENERGY CONVERSION BY ELECTROCHEMICAL MEANS

Electrochemical systems are classified into three categories, viz:

(i) Substance producer (ii) Energy producer and (iii) The substance and energy destroyer.

The energy producer is a spontaneously working self-driving electrochemical system (Fig.1.4). In it, there is a spontaneous occurrence of a dé-electronation reaction at the electron sink electrode and an electronation reaction at the electron-source electrode. If an external load is connected to the two electrodes, a current of electrons flows in the external circuit. Thus the electrodic reaction at the two electrodes bring about the conversion of the difference in the chemical energy (Gibb's free energy) of the reactants and the products directly into a flow of electricity (and thereafter by a motor, to mechanical energy).

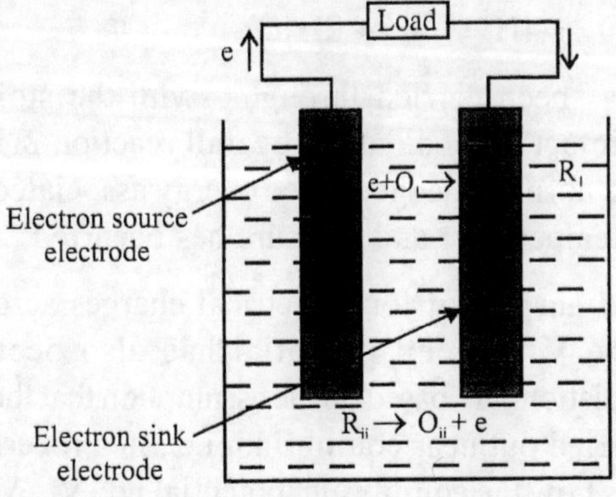

Fig.15.4 Schematic diagram of an electrochemical energy producer

There is no intermediate step in which the energy has to bring itself to power by expansion of a gas converting thereby only a part of its thermal energy to mechanical work.

THE MAXIMUM INTRINSIC EFFICIENCY IN ELECTROCHEMICAL CONVERSION OF THE ENERGY OF A CHEMICAL REACTION TO ELECTRIC ENERGY

We have in Thermodynamics the important relation:

$$-\Delta G = W_{rev} - P\Delta V \qquad \qquad .....(15.2)$$

which means that the change in free energy in a reaction is equal to the *total reversible work* obtainable from the reaction (this work to include all kinds of work, i.e., gravitational, electrical,surface, etc.,and also the work of expansion) diminished by the work of expansion, $P\Delta V$. Hence,

$$-\Delta G = W'_{rev} - P\Delta V \qquad \qquad .....(15.3)$$

where W'_{rev} is all the work obtainable from the reaction exclusive of any work which can be obtained from a possible volume change in the system.

Now if one carries out a chemical reaction in an electrochemical way, then in the example of the reaction $2H_2 + O_2 \rightarrow 2H_2O$, there will be two partial reactions

$$2H_2 \rightarrow 4H^+ + 4e \qquad \qquad .....(15.4)$$

and $\qquad O_2 + 4H^+ + 4e \rightarrow 2H_2O \qquad \qquad .....(15.5)$

After each has been carried through with the stoichimetric quantities, one has infact carried out the overall reaction $2H_2 + O_2 \rightarrow 2H_2O$ and hence, the normal change of free energy associated with this reaction at a given temperature and pressure has occurred.

But also, the transport of four electrical charges across a total potential difference V, the cell potential has also occurred. As thermodynamic calculations are based on the assumption that the chemical change has been carried out near equilibrium i.e., in a reversible way, the cell potential V is now the equilibrium potential i.e., $V = V_e$. It is that obtained on an electronic voltmeter with the velocity of the electrode reactions at "infinitely slow" in conformity with the conditions of thermodynamic reversibility. Now the electrical work of transporting such charges (four electrons per two molecules of water formed, or 4 Faradays, if molar quantities are considered) is the total charge transported multiplied by the potential difference through which it passes i.e.,$4FV_e$. Thus the general expression for the change of free energy in one act of an electrochemical reaction in which the number of electrons transported externally for each act of the equivalent elctrode reactions is 'n', is

$$nFV_e \qquad \qquad .....(15.6)$$

Compare this *electrical* work carried out in the reaction, now with W'_{rev}, the total work obtainable from the reaction excluding volume-change work. What other kind of work than electrical work is obtainable from the reaction $2H_2 + O_2 \rightarrow 2H_2O$ *carried out in this electrochemical way?* There is no surface work and no gravitational work. Carried out in this electrochemical way and in an ideal manner (i.e., infinitely slowly so that the potential differences in the cell are those characterstic of equilibrium),

$$W'_{rev} = nFV_e \qquad \qquad .....(15.7)$$

Hence from eq. (15.3),

$$-\Delta G = nFV_e \qquad \qquad .....(1.8)$$

In Table15.1 are listed the Gibbs free-energy change and the corresponding equilibrium potential difference for the reactions of the oxidation of some currently used and potential fuels.

Table 15.1 Theoretical Cell Potentials of various oxidation Reactions

Fuel	Reaction	$\ddot{A}G^o$ k.cal.mole^{-1}	V_e^0,V
Hydrogen	$H_2 + \frac{1}{2} O_2 \rightarrow H_2O$	-56.69	1.229
	$H_2 + Cl_2 \rightarrow 2HCl$	-62.70	1.370
Propane	$C_3H_8 + 5O_2 \rightarrow 3CO_2 + 4 H_2O$	−503.90	1.093
Methane	$CH_4 + 2O_2 \rightarrow CO_2 + 2H_2O$	-195.50	1.060
Carbon monoxide	$CO + \frac{1}{2}O_2 \rightarrow CO_2$	-61.45	1..333
Methanol	$CH_3OH + {}^3/^2O_2 \rightarrow CO_2 + 2H_2O$	-168.95	1.222

Contd...

Fuel	Reaction	$\ddot{A}G^0$ k.cal.mole^{-1}	V_e^0,V
Formaldehyde	$CH_2O + O_2 \rightarrow CO_2 + H_2O$	-124.7	1.350
Hydrazine	$N_2H_4 + O_2 \rightarrow N_2 + 2H_2O$	-143.9	1.560
Zinc	$Zn + \frac{1}{2}O_2 \rightarrow ZnO$	-76.05	1.650
Sodium	$Na + \frac{1}{2}H_2O + 1/4O_2 \rightarrow NaOH$	-71.84	3.120
Carbon	$C + O_2 \rightarrow CO_2$	-94.26	1.020

It is in this sense, that in an electrochemical energy converter, the ideal maximum efficiency is 100% for, as in the above idealized situation, if one could carry out reactions in such a way that the electrode potentials were infinitely near the equilibrium values, the electrical energy one would draw from the reaction would be nFV_e and this is all of the free energy change ΔG, which is the maximum amount of useful work one can obtain from a chemical reaction. Thus what has been realized is that the Intrinsic maximum efficiency of the electrochemical converter workingr under ideal conditions is 100% of the ΔG, which is the useful or intrinsically *available* work of a chemical reaction.

It has been shown that the electrochemical method of conversion to electricity all the energy which is intrinsically available in a chemical reaction (independently of the method of conversion) is of maximum efficiency. Not quite all the energy difference between the reactants and products of an electrochemical reaction can be made available, however even by electrochemical method because some of it is wasted in very fundamental processes connected with entropy losses and entropy gains which also occur in chemical reactions. It is the enthalpy change ΔH which is equivalent to the total change in energy between the reactants and products of a reaction *including* the energy lost in entropy increases. It is a more significant standard of comparison, to base the efficiency of any energy conversion method on a comparison of how much energy it gives compared with heat content(enthalpy) change, ΔH, in a reaction because $\ddot{A}H$ is the total energy difference between products and reactants of a reaction. The $\ddot{A}H$ is usually larger in magnitude than ΔG often by 10 to 20%. Hence a better expression for the intrinsic maximum efficiency of an ideal electrochemical converter is,

$$\varepsilon_{max} = \frac{\Delta G}{\Delta H} = -\frac{nFV_e}{\Delta H} \qquad .....(15.8)$$

The cell potentials depend on whether the reaction is carried out in acid or alkaline electrolyte, since the latter case, the reaction product is a carbonate with a somewhat different standard free energy than CO_2. Since the activity of the reactants and products depends on the concentration of the electrolyte, so does the potential of the cell reactions and the efficiency of conversion. Still, the maximum intrinsic efficiency for electrochemical energy conversion on a heat- content comparison basis is in the region of 90% compared with 20-40% of heat engines operating under tolerable temperature linits.

THE ACTUAL EFFICIENCY OF AN ELECTROCHEMICAL ENERGY CONVERTER

It has been shown that in an electrochemical energy converter, the maximum cell potential is the value V_e obtainable when the reaction in the cell is electrically balanced out to equilibrium, i.e., when no current is being dawn from the cell. *As soon as the cell drives a current through the external circuit, the cell potential falls from the equilibrium value V_e to V.* The value of the actual potential V at which the cell works when delivering a current 'i' is always *less* than the equilibrium potential V_e. Hence eq (15.8) assumes the form

$$\varepsilon_0 = -\frac{nFV_e}{\Delta H}\frac{V}{V_e} \qquad \qquad .....(15.9)$$

or

$$\varepsilon_0 = \varepsilon_{max}\,\varepsilon_p \qquad \qquad .....(15.10)$$

where ε_{max} is the maximum efficiency given by eq. (15.8), and ε_p is known as the *voltage efficiency* given by

$$\varepsilon p = \frac{V}{V_e} \qquad \qquad .....(15.11)$$

Of course this picture is true only if the reactants are completely converted into products, i.e., the overall reaction is fully accomplished and none of the electrons take part in some alternative reaction.

To allow for the possibility that such a wastage does occur, we must consider the current or faradic efficiency ε_f to take into account the incomplete conversion of reactants into products. The overall efficiency will be

$$\varepsilon_o = (\varepsilon_{max} \, \varepsilon p) \, \varepsilon_f \qquad \qquad .....(15.12)$$

In many reactions of interest, ε_f is virtually unity.

In electrochemical converter, there are no collisions between the particles (e.g., H_2 and O_2) which react to form, e.g., H_2O. They simply undergo electron -supplying (for the H_2) and for electron accepting (for the O_2) reactions on the electrodes. These electrons travel round an external circuit and,while doing so, pass through a *load* (e.g,, the armature of a motor) and thereby do work. They complete the reaction and allow the product water to be formed. The electrons pass through the *entire* potential difference generated by the cell reaction, and not through a part of it only.

Why then,the electrochemical converter working under ideal conditions does not convert into electric energy all the energy released in a chemical reaction ΔH, but only the free–energy change in the reaction, ΔF? This is because the reaction taking place is still the same overall chemical reaction as in thermal reaction, it is still $2H_2 + O_2 \rightarrow 2H_2O$.

The *entropy* change will be unaltered. A part of the ΔH is used up in the unavailable energy connected with differences in order and disorder between the products and reactants and this is independent of any method of energy conversion.

COLD COMBUSTION

The net cell reaction (the summation of the two electrode reactions $2H_2 \rightarrow 4H^+ + 4e$ and $O_2 + 4H^+ + 4e \rightarrow 2H_2O$ in which the electrons cancel out) is equal to the actual combustion reaction; it gives out heat which may be converted to mechanical work with only an efficiency given by the Carnot expression eq. (15.1) . The rest of the heat is evolved as heat. But, in the electrochemical reaction the heat ΔH is not given out.

What is given out is not hotter molecules (made hotter by transfer of the difference of the potential energy of the rectants and products to kinetic energy) but a stream of electrons, the total energy of which is ΔG (per Avogadro number of the act of the overall reaction). The combustion reaction has occurred but *cold.*

It is seen that the same reaction has occurred as in combustion and the free energy ΔG has been drawn off. The total energy change in the reaction, however, is ΔH. Hence there is some heat energy given up or taken in during reversible working of an electrochemical reactor. It is $\Delta H - \Delta G$ or $T\Delta S$ and is usually negative, i.e., heat is given out. The amounts are small, e.g., in the oxidation of one mole of propane into carbon dioxide, $\Delta H = -530.6$ Kcal.mole^{-1}, but $T\Delta S$ is only -32 Kcal.mole^{-1}. Electrochemical combustion is *almost* cold.

THE ELECTROCHEMICAL ENGINE

The principal use hitherto of electrochemical energy converter is *auxiliary power in space* in which the *weight* of energy converter plus fuel carried is prime importance. In any energy conversion situation where transportation, in its most general sense, the system must carry its own fuel with it to provide a certain number of kilowatt hours of energy during a journey of known duration, the main point is the *efficiency* of conversion; the weight of the fuel necessary for a given operation is clearly inversely proportional to it. Hence the use in space exploited the avoidance of Carnot cycle efficiency loss in electrochemical conversion.

Another probable use of electrochemical reactor, however, is as a *power* (rather than energy) source and, in combination with an electrical motor, as an electrochemical engine. Here the stress is on power, the rate of delivery of energy per unit weight, and not so much on the efficiency of conversion, though this is also important. There is at present no advantage to be gained from an electrochemical engine over a thermal combustion engine in respect of a power to weight ratio although they are better than diesel engines in this respect.

THE POWER OUTPUT OF AN ELECTROCHEMICAL ENERGY CONVERTER

When electrochemical energy converters are to be considered for use, e.g., in driving trucks, it is the power output or the rate at which they are able to do work that has to be examined. The power P of an electrochemical energy converter is defined thus

$$P = IV \tag{15.13}$$

From which it follows that the power is small when I is small although V is at its maximum V_e. But the power output is also small when I is very large because of the sudden growth of concentration over potential when the current density approaches i_L tends to drive V to zero. Thus, the P versus I curve should pass through a maximum Fig. 15.5. The distinction between high efficiency and high power will be evident if one compares the P versus I and ε versus I curves

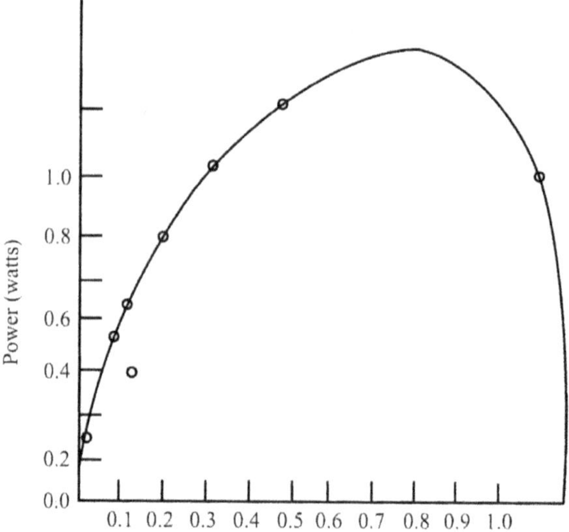

Fig. 15.5 *The power versus current-density relations for an electrochemical energy converter*

Fig. 15.6. When its efficiency is at a maximum, the electrochemical energy converter is a less good power source. As the current density is increased, the power output increases, but the efficiency decreases. But ofcourse, at the highest power drains, both the power and efficiency fall towards zero.

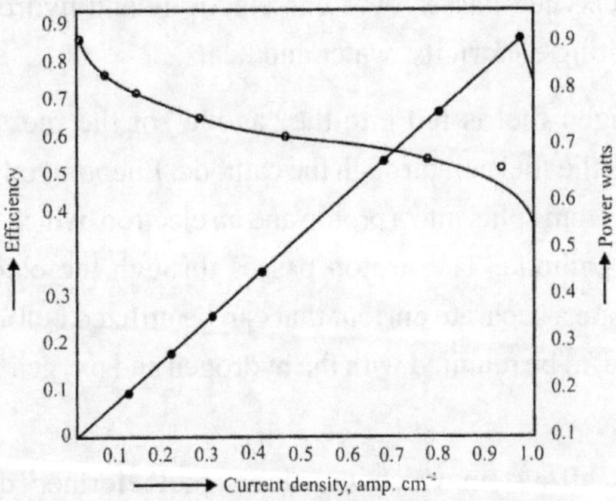

Fig. 15.6 The power versus current density and efficiency versus current density relations for the electrochemical energy converter

Fuel Cells

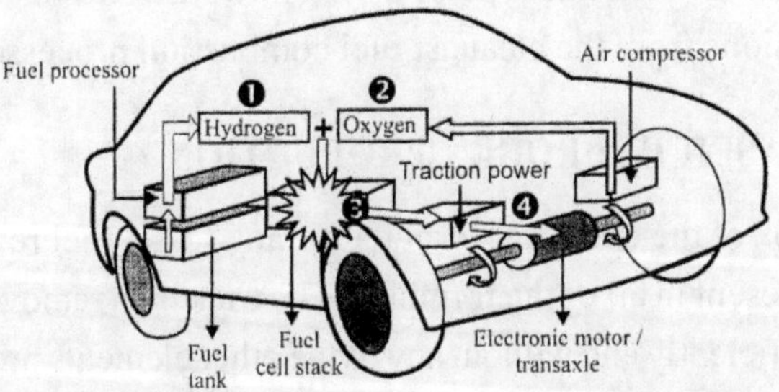

❶ The fuel processor converts methanol (or gasoline) to hydrogen. Hydrogen is supplied to the fuel cell

❷ Air is supplied to the fuel cell by the air compressor.

❸ Oxygen from the air and hydrogen from the fuel processor combine in the fuel cell to generate electricity, which is sent to the traction inverter module.

❹ The traction inverter module converts the electricity for use by the motor/transaxle. The motor / transaxle converts the electric energy into the mechanical energy, which turns the wheels.

Chemistry of a Fuel Cell

Anode side : $2H_2 => 4H^+ + 4e^-$ Cathode side : $O_2 + 4H^+ + 4e^- => 2H_2O$

Net reaction : $2H_2 + O_2 => 2H_2O$

In principle, a fuel cell operates like a battery. Unlike a battery, a fuel cell does not run down or require recharging. It will produce energy in the form of electricity and heat as long as fuel is supplied.

A fuel cell consists of two electrodes sandwiched around an electrolyte. Oxygen passes over one electrode and hydrogen over the other, generating electricity, water and heat.

Hydrogen fuel is fed into the "anode" of the fuel cell. Oxygen (or air) enters the fuel cell through the cathode. Encouraged by a catalyst, the hydrogen atom splits into a proton and an electron, which take different paths to the cathode. The proton passes through the electrolyte. The electrons create a separate current that can be utilized before they return to the cathode, to be reunited with the hydrogen and oxygen in a molecule of water.

A fuel cell system which includes a "fuel reformer" can utilize the hydrogen from any hydrocarbon fuel – from natural gas to methanol, and even gasoline. Since the fuel cell relies on chemistry and not combustion, emissions from this type of a system would still be much smaller than emissions from the cleanest fuel combustion processes.

FOSSIL FUEL BASED HYDROGEN PRODUCTION

A closer look at the chemical formula for any fossil fuel reveals that hydrogen is present in all of the formulas. The trick is to remove the hydrogen safely, efficiently and without any of the other elements present in the original compound. Hydrogen has been produced from coal, gasoline, methanol, natural gas and any other fossil fuel currently available. Some fossil fuels have a high hydrogen to oxygen ratio making them better candidates for the reforming process. The more hydrogen present and the fewer extraneous compounds make the reforming process simpler and more efficient. The fossil fuel that has the best hydrogen to carbon ratio is natural gas or methane (CH_4).

STEAM REFORMING OF NATURAL GAS

Hydrogen

Photoelectrolysis, known as the hydrogen holy grail in some circles, is the direct conversion of sunlight into electricity. Photovoltaics, semiconductors and an electrolyzer are combined to create a device that generates hydrogen. The photoelectrolyzer is placed in water and when exposed to sunlight begins to generate hydrogen. The photovoltaics and the semiconductor combine to generate enough electricity from the sunlight to power the electrolyzer. The hydrogen is then collected and stored. Much of the research in this field takes place in Golden, Colorado at the National Renewable Energy Laboratory

Types of fuel cells: Solid oxide; Molten carbonate; Phosphoric acid; Solid polymer; Direct methanol; Alkaline

Fuel cells are energy conversion devices that transform the energy stored in a fuel into electricity and heat. The fuel is not burned in a flame but oxidised electrochemically. This means that fuel cells are not constrained by the fundamental law that governs heat engines, the so-called Carnot limit, which specifies the maximum theoretical efficiency that a heat engine can reach. Fuel cells have much in common with batteries, which also convert energy that is stored in chemical form into electricity. In contrast to batteries, however, they oxidise externally supplied fuel and therefore do not have to be recharged.

Fuel cells have been invented in 1839 but have not been widely applied because the technology is demanding, and because other technologies have been dominating the marketplace. However, their exceptional environmental performance and their outstanding efficiencies have recently caused renewed interest, buoyed by advances in engineering and material science. Fuel cells may today provide a real alternative for a cleaner way of handling power. At the heart of any fuel cell is the electrolyte which separates the two electrodes. There are several different types of electrolytes with very different properties, and hence very different fuel cell types have been built around them, mostly named after the electrolyte. They are:

SOLID OXIDE FUEL CELLS (SOFC)

A SOFC uses yttria-stabilised zirconia as its electrolyte, sandwiched between the anode and the cathode. It runs at a temperature of around 1,000 °C. The heat produced can be used in cogeneration applications or in a steam turbine to provide more electricity than that generated from the chemical reaction within the fuel cell (a bottoming cycle). A number of different fuels can be used, from pure hydrogen to methane to carbon monoxide, and the nature of the emissions from the fuel cell will vary correspondingly with the fuel mix.

There are three fundamental designs of SOFC – the tubular, planar and monolithic types. The first of these was designed by the Westinghouse Electric Corporation and operates with the fuel on the outside surfaces of a bundle of tubes, and the oxidant on the inside, the tube itself being composed of the electrolyte and electrode 'sandwich'.

Planar SOFCs are under development by a number of companies, with Siemens and Fuji Electric two of the leaders. In this case the cells are flat plates bonded together and placed one on top of the other to form a stack. The advantages of this system over the tubular system are its relative ease of manufacture and a lower ohmic resistance of the electrolyte, resulting in reduced energy losses.

Monolithic SOFCs are in a very early stage of design, with the process one of sintering and corrugation of the electrodes and electrolyte to form a honeycomb structure. Basic laboratory tests have been conducted, with results indicating that this form of fuel cell may be one of the most efficient.

MOLTEN CARBONATE FUEL CELLS (MCFC)

MCFCs are so named because the electrolyte they use is a molten alkali carbonate mixture, retained in a matrix. They operate at a temperature of about 650 °C, meaning that once again useful heat is produced. In this case the cathode must be supplied with carbon dioxide, which reacts with the oxygen and electrons to form carbonate ions, which carry the ionic current through the electrolyte. At the anode these ions are consumed in the oxidation of hydrogen, which also forms water vapour and carbon dioxide to be transferred back to the cathode. There are two ways of doing this: either by burning the anode exhaust with excess air and removing the water vapour before mixing it with the cathode inlet gas; or by separating the CO_2 from the exhaust gas using a 'product exchange device'.

The fuel consumed in an MCFC is usually natural gas, though this must be reformed in some way to create a hydrogen-rich gas to feed to the stack. An MCFC produces heat and water vapour at the anode, which can be used for the steam reformation of methane. This means that it is inherently more efficient than a cell requiring external fuel processing. Again, the MCFC can use carbon monoxide at the anode as a fuel.

The MCFC is seen by many as an ideal source for large scale power generation. One reason for this is the necessity for large amounts of ancillary equipment, which would render a small operation uneconomic. There is also no requirement for expensive catalysts as in low temperature fuel cells, and a third reason is the fact that the heat generated can be used for internal reformation of methane, a bottoming cycle and for fuel processing and cogeneration. This increases the efficiency of the fuel cell system.

PHOSPHORIC ACID FUEL CELLS (PAFC)

The PAFC is one of the oldest and therefore most established fuel cell technologies, and is the only one in use in a small number of power generation projects. It uses phosphoric acid as its electrolyte, and is able to reform methane to a hydrogen-rich gas for use as a fuel with the waste heat from the fuel cell stack. This heat may also be utilised for space heating or hot water. It operates at temperatures around 200 °C, so the waste heat is not of a high enough quality to be used in cogeneration applications. It is also possible to use alcohols such as methanol and ethanol as fuels, though care must be taken in all cases to avoid poisoning the anode by carbon monoxide and hydrogen sulphide which may be present in the reformed fuels. This results in a gradual reduction in performance and the eventual failure of the cell.

The Japanese are particularly advanced in PAFC research and design, with power plants from a few kilowatts to a few megawatts already in operation. The major keiretsu are all in evidence, among them Toshiba, Fuji and Mitsubishi, as they see Japan's lack of natural resources one way of forcing the technology upon the market at a higher price than would be possible in many other countries.

SOLID POLYMER FUEL CELLS (SPFC)

The SPFC (also known as the proton exchange membrane fuel cell, PEMFC) is unusual in that its electrolyte consists of a layer of solid polymer which allows protons to be transmitted from one face to the other. It basically requires hydrogen and oxygen as its inputs, though the oxidant may also be ambient air, and these gases must be humidified. It operates at a temperature much lower than the fuel cells mentioned so far, because of the limitations imposed by the thermal properties of the membrane itself. The operating temperatures are around 90 °C. The SPFC can be contaminated by carbon monoxide, reducing the performance by several percent for contaminant in the fuel in ranges of tens of percent. It requires cooling and management of the exhaust water in order to function properly.

There are a number of companies involved in manufacturing SPFCs. Ballard are probably the leaders, though companies such as DeNora in Italy and Siemens are progressing fast. The main focus of current designs is transport applications, as there are advantages to having a solid electrolyte for safety, and the heat produced by the fuel cell is not adequate for any form of cogeneration. Daimler-Benz has taken a high profile in developing cars powered by Ballard fuel cells, while Toyota has recently presented a vehicle that is using a fuel cell of their own design. Other car manufacturers, including General Motors and Ford, are actively engaged in similar developments. It now appears, however, that there is a strong possibility of using the SPFC in very small scale localised power generation, where the heat could be used for hot water or space heating. There is also the possibility of a heater/chiller unit for cooling in areas where air conditioning is popular. If it does prove possible to use this particular type of fuel cell for both transport and power generation, then the advantages generated by economies of scale and synergy between the two markets could make the introduction of the technology easier than in other cases.

DIRECT METHANOL FUEL CELLS (DMFC)

This type of fuel cell is based on solid polymer technology but uses methanol directly as a fuel. If it can be made to work, that would be a big step forward in the automotive area where the storage or generation of hydrogen is one of the big obstacles for the introduction of fuel cells. Prototypes exist, but the development is at an early stage. There are principal problems, including the lower electrochemical activity of the methanol as compared to hydrogen, giving rise to lower cell voltages and, hence, efficiencies. Also, methanol is miscible in water, so some of it is liable to cross the water-saturated membrane and cause corrosion and exhaust gas problems on the cathode side. Nevertheless, the direct methanol fuel cell is an interesting proposition and a number of places are working on it, including Siemens in Germany, the University of Newcastle and Argonne National Laboratory.

There are also efforts to develop a low-temperature SOFC (500 °C) that would also allow the use methanol directly, as well as using stainless steel components. The idea is still young but intriguing. Imperial College in London is active in this area.

ALKALINE FUEL CELLS (AFC)

Alkaline fuel cells use a solution of potassium hydroxide in water as their electrolyte, making them sensitive not to CO as the SPFC is, but to CO_2, meaning that oxygen has traditionally been used as the oxidant in the system. This has led to few uses outside aerospace, although some of the first experimental vehicles were powered by AFCs. They use comparatively cheap materials in their electrodes but are not as power dense as SPFCs, making them bulky in some situations.

The alkaline fuel cell has been used with great success in the past in space missions, dating back to the Apollo and Gemini missions in the 1960s. It is still in use in the Space Shuttle today and provides not only the power but also the drinking water for the astronauts. The space-qualified hardware made by United Technologies Corporation is very expensive, and until recently it was not thought that it would be useful in other applications. However, in July 1998 the Zero Emission Vehicle Company (ZEVCO) launched its first prototype London taxi based on the technology. Using a 5 kW fuel cell and about 70 kW of batteries, the hybrid taxi is a zero emissions vehicle capable of operating in city centres, though some refinements are needed before the final version hits the roads. The AFC, like the SPFC, is happiest when fed pure hydrogen, so the issue of refuelling must be addressed if these vehicles are going to progress in the marketplace.

Questions

1. Discuss the principle of a thermo ionic conductor for generation of electricity.

2. Discuss the principle of operation of a Magneto Hydro dynamic converter.

3. What is a fuel cell ? Describe the principle of the functioning using Hydrogen and Oxygen.

4. What are the advantages and discuss advantages of the fuel cell ?

5. Discuss the application of the tool cell in motor transport sector.

16

ENERGY SCENERIO - INDIA

India has a population of about 97.97 crores distributed in 26 states. The country uses 12.18 Quads of energy as against a production of 9 Quads, which is about 3% of the world's total energy. For a long time the people of India were using fuel from rural resources such as wood, dried animal dung, husk and the like towards generation of energy at homes. Even heavy industrialisation did not take place at that time , that is upto the year 1952. Owing to rapid industrialisation and sophisticated home equipment, the power needs have gone up. The resources for power generation in India are: coal (55%), petroleum (30%), natural gas (7%), hydroelectric (3%), nuclear (1.3%) and the remainder from forest and rural sources. However these resources today do not provide even a thirtieth fraction of what the advanced country like USA is consuming for the energy needs; in fact the per capita consumption of energy in India is even one fifth of the world's average.

STATUS OF FOSSIL FUELS

The country's coal production and consumption are almost balanced. India has vast proven resources of coal in 5 states; almost 65% of it goes in the electricity generation. The country is the third largest producer of coal next to China and USA. Major portion of the coal is not superior grade and hence high quality anthracite coal is imported from Australia and New Zealand.

India's oil production runs on deficit as compared to the consumption; about 747 thousand billion barrels per day produced and 1930 thousand billion barrels per day consumed. Therefore oil is imported from Iran, Iraq, Saudi Arabia and Oman to a large extent. The oil wells were dug in Bombay High, Krishna-Godavari-Kaveri basins and Assam fields. The indigenous oil together with the imported crudes are refined at the refineries in India (Table 16.1). Currently there are plans to expand both drilling and refining activities to cope up with the industrial development. Approcximately 30% of oil goes into power generation.

Table 16.1 *Petroleum refineries in India*

Name of Refinery	Location	Capacity, bbl/day
Indian Oil Corporation Ltd (IOL)	Gauhathi	20,000
	Barauni	84,000
	Koyali	251,000
	Haldia	75,000
	Mathura	150,000
	Digboi	13,000
	Panipat	120,000
Hindusthan Petroleum Corporation Ltd (HPCL)	Mumbai	110,000
	Visakapatnam	90,000
Bharat Petroleum Corporation Ltd (BPCL)	Mumbai	138,000
Madras Refineries Ltd (MRL)	Chennai	130,000
Cochin Refineries Ltd (CRL)	Cochin	150,000

Table 16.1 Contd...

Name of Refinery	Location	Capacity, bbl/day
Bongaigaon Refinery & Petrochemicals Ltd (BRPL)	Bongaigaon	47,000
Crude Distillation Unit of MRL	Narimanam	10,000
Numaligarh Refineries Ltd (NRL)	Numaligarh	60,000
Mangalore Refinery & Petrochemicals Ltd (MRPL)	Mangalore	195,000
Reliance Petroleum Ltd (RPL)	Jamnagar	542,000
Total		2,185,000

India's natural gas production and consumption are quite balanced and the figures are 0.75 and 0.752 trillion cubic feet per year. The major gas sources are Bombay High, Heera, Panna, S.Bassein, and Neelam wells situated in the west coast. The country has plans to construct 12 LNG terminals to supply the gas to various installations in the west and east coasts and when these ventures are completed, the need to import natural gas to the tune of at least 50 million metric tons /year arises.

NUCLEAR ELECTRICITY

India produces from its 14 nuclear power plants a total of 2720 Mwe electricity. A list of important plants is given in Table12.2. Lack of fuel resources, disposal facilities, funds and technologies are the impediments in harnessing this power. A 13 MWe fast breeder plant is operating at the Indira Gandhi Center at Kalpakkam since 1997 and two more are being planned.

HYDROELECTRICITY

Hydroelectricity generation in India has a great potential with 86000 MWe as against the installation of about 22000 MWe. Indian engineers have rich experience in constructing dams for power generation since early 1930s. Some major hydroelectric power plants in India are listed in Table 16.2. The tapping of hydroelectricity even from small reservoirs and water falls is taking rapid strides in the development of this energy in the country.

Table 16.2 *Major hydroelectric power plants in India*

Name of Power Plant	State	Capacity, MW
Dehar	Rajasthan	990
Sharavathi	Karnataka	891
Koyna	Maharashtra	880
Kalinadi-1	Karnataka	825
Nagarjuna Sagar	Andhra pradesh	815
Idukki	Kerala	780
Srisailam Right Bank	Andhra pradesh	770
Bhakra-Nangal	Rajasthan	710
Salal	Jammu & Kashmir	690
Kundah	Tamilnadu	555

SOLAR ENERGY

India receives almost 300 clear sunny days in a year with an average daily energy of about 4-7 kW/ m^2. The resource is expected to yield over 5000 trillion kWh/year which is sumptuous. The need for exploiting this natural resource was recognised as early as 1975 and the government encouraged a number of projects towards harnessing solar energy. Till date about 800000 gadgets namely, lights, cookers, water/fluid pumps and power packs and grid connecting systems were installed in the country. Considering the magnitude of the resource and the funding this progress is not commensurate and a great deal of progress has to be achieved.

WIND POWER

The potential for tapping wind power in India is to the extent of 20000 MW against which only about 1175 MW was exploited. The country's West Coast receives steady winds for the periods April to September of the year and this season is most suitable for power generation. The commercialisation of this energy lagged behind due to some uncertainties such as weather, high cost equipment, maintenance and users repulsions.

GEOTHERMAL ENERGY

The geothermal energy production share of India is practically nil on account of many tardy locations from where it can be tapped. The Geological Survey of India identified 340 potential sites, of which 11 were to render the energy at depths of 1-3 km (water temperatures 100-120 °C). The three most proven sites in India are:

NW Himalayas

Puga-Chumatang (Ladakh district, J &K)

→ 1 MWe plant planned.

Parbati Valley, Manikram field (Himachal Pradesh)

→ 5 kWe binary cycle plant functioning since 1992.

Central India: Tattapani region (Madhya pradesh)

→ 20 MWe binary plant planned.

The progress in the recovery of geothermal energy could not go in pace with hydroelectric energy generation on account of uncertain locations, high costs and low yields.

ENERGY FROM BIOMASS

For a country like India which depends on agriculture biomass is the best suited energy resource, because it is abundant. Moreover urban areas in India contribute to sufficient quantity through garbage. If the biomass is harnessed for energy in rural and urban areas it will be a worthwhile proposition, and this option must be thoroughly examined by the policy makers. There is also a need to recycle some of the useful materials emanating from the urban wastes. This will not only improve the economy but also solve the power crisis problems of the country.

ELECTRICITY STATUS

India's net generation of electricity in the year 1999 stands at 424 billion kilowatt hours. Out of this, at least 21% of power is lost due to outages and misuse. The consumption during the same year is 396 billion kilo watt hours.

The country faces an acute shortage of 11% during normal times and 18% in peak summer. The total installed electricity was 103,445 MWe in 1999. About 75% of it is produced from thermal power plants, 22% from hydroelectricity generators, 2% from nuclear power plants and the remainder from other sources. A list of the large thermal power plants installed in India is given in Table 16.3.

Table 16.3 Large Thermal Power Plants in India (Fuel type)

Power Plant	Capacity, MW	Fuel *	State
Korba West	2100	Coal	Madhya Pradesh
Ramagundam	2100	-do-	Andhra Pradesh
Neyveli	2070	-do-	Tamilnadu
Singrauli	2050	-do-	Uttar Pradesh
Chandrapur	1900	-do-	Maharashtra
Vindhyachal	1760	-do-	Madhya Pradesh
Anpara	1630	-do-	Uttar Pradesh
Farakka	1630	-do-	West Bengal
Obra	1550	-do-	Uttar Pradesh
Talcher	1470	-do-	Orissa
Uran	672	Gas	Maharashtra
Trombay	625	-do-	""
Dadri	524	-do-	Uttar Pradesh
Paguthan	470	-do-	Gujarat
Auraiya	448	-do-	Uttar Pradesh
Gandhar	393	-do-	Gujarat
Hazira Essar	350	-do-	""
Baroda GIPCL	312	-do-	""
Anta	264	-do-	Rajasthan
Kayamkulam	230	-do-	Kerala
Dabhol	453	Naphtha	Maharashtra
Kawas	424	-do-	Gujarat
Chennai Vasavi	206	Oil	Tamilnadu
S. Bassein	152	-do-	Maharashtra
Pampore	140	-do-	Jammu & Kashmir
Jamnagar RIL	132	Naphtha	Gujarat
Yelahanka	131	Oil	Karnataka
Basin Bridge	124	Naphtha	Tamilnadu
Brahmapuram	110	Oil	Kerala
Maharashtra cracker	79	-do-	Maharashtra

In addition many new ventures for power generation were planned and are being executed in recent years. A total power of about 8000 MWe is expected to be added in the coming years from the plants listed in Table 16.4.

Table 16.4 Large Thermal Power Plants under construction in India

Power Plant	Capacity MW	Fuel Choice	State	Status of Completion
Korba East 1 &2	500	Coal	Madhya Pradesh	2003
Dabhol 2 GT 1 &2	500	Gas	Maharashtra	2001
Dabhol 3 GT 1&2	500	-do-	-do-	-do-
Dabhol 2 Sc1	300	Waste heat	-do-	-do-
Dabhol 3 Sc 1	300	-do-	-do-	-do-
Bhatinda refinery	500	Oil	Punjab	N/a
North Madras 3 No.1&2	526	Coal	Tamilnadu	2001
Pillaiperumalnallur	330	Naphtha	-do-	2002
Bakreshwar 1 & 3	630	Coal	West Bengal	2001
Bakreshwar 1 & 2	500	-do-	-do-	2002

ENVIRONMENT STATUS

As common in any developing country, India faces environmental pollution due to expansion of industry, increased number of thermal power plants (mostly using coal), increased number of automobiles on roads, large domestic wastes dropped by excessive population and enhanced greenhouse gases. The shortage of rainfall, the sinking levels of groundwater and lack of anti-pollution technologies are adding up to the risks of environment. The three Indian cities that are highly polluted according to statistics given by WHO are New Delhi, Mumbai and Kanpur. A large number of research laboratories have been studying the problems of environmental pollution that is caused by the man-made events and rapid solutions were evolved. Environmental Impact Assessment studies have been conducted on all the major and minor industries in existence including the expansion phases of the industries.

The Central and State Pollution Control Boards enforce two laws namely, The Water (Prevention and Control of Pollution) Act, 1977 and The Air (Prevention and Control of Pollution) Act, 1981 on the Public and the Industry. On the whole, the situation as regards environmental damage is not very alarming as remedial actions are taken quickly on time.

AN OVERVIEW

Presently there are many pitfalls in the generation, distribution and management of energy in India. Even under the best managements at both centre and state levels the deficits of energy requirements could not be filled. The electricity production as well as consumption has been steadily increasing by 10% every year starting from 1988 and the situation is *status quo*. To start with, it appeared that individual power producers may play a key role in the generation and distribution of electricity and with this view the State Electricity Boards have been constituted throughout the country. Apparently all the Boards have been 'fund starved' and were showing losses that could not be easily made up. Moreover, energy losses in any of the Indian States range from 20-22% causing several hardships for the managements to progress. In addition, the annual growth in production is not significantly high enough to compensate for the losses.Under these circumstances, it is opined that the following measures , if implemented, may increase the productivity in energy management of the country.

- All the State Electricity Boards are to be revamped and made to earn profits and increase the productivity

- Annual increase of electricity production shall be enhanced to cover up only transmission losses in the lines and not mismanagement or stealth of power from the lines.

- The current deficit in power supply shall be brought to zero level

- National priorities shall be changed to see that the power sector gets boost, as this alone can bring progress and employment.

- All regional energy resources, however small in magnitude they may be, are to be harnessed and the energy supplied locally or added to the grid. In this connection the following renewable energy resources need to be further exploited.

SOLAR ENERGY

The entire country has this energy resource. It is not just enough to manufacture a handful of gadgets and propogate their sale to the people, but the effort shall be to establish 'solar power plants' that can really augment the regular energy supplied through the grid. Solar power stations thus established in desert regions can be utilised to produce 'hydrogen', a reserve of energy that can be converted back to energy at will. Technological development in this direction would yield better productivity to the nation.

HYDRO POWER

The statistics reveal that there is a great potential to install more and more power generation units. Almost all the states in India have the moving water falls, high or low. Every natural setting that provides hydro power has to be harnessed.

BIOFUELS

India should focus its attention on the development of biofuels from the biomass that is abundant. These materials can be manufactured at low cost even in rural areas and hence this gives a boost to village industries. Even cultivation of high energy yielding crops shall be encouraged to a reasonable level. Within the urban front, the biomass after separation from recyclable mass should be subjected towards the generation of electrical power rather than allowing it to submerge on earth.

India's renewable energy potential and achievements are recorded in the TERI Yearbook- 2002, which has been shown in Appendix-1.

It is thus possible to achieve self-sufficiency in the power sector if all these measures are ensured by the policy makers and managers who frame the destiny of the country. As regards the environmental pollution, the levels are not so alarming when compared to industrial world, but constant surveillance has to be kept.

17

ENERGY AUDIT

Energy audit is very much similar to environmental audit in an industry. The use of energy has to be critically accounted and any saving if possible shall be achieved. This is possible by adopting suitable pre-audit testing, technical changes and introduction of new policies. In short, a few of the environmental audits serve as guidelines for energy audit as well; thus energy audit is a *'green technology'* incorporating within its periphery the technical fixes, clean technologies and sustainable energy. The scope of this audit is expected to bring in potential savings in the industrial sectors listed below:

1. Electrical lamp efficiency enhancement
2. Industrial & Domestic sector space heating alterations
3. Domestic cookers efficiency
4. Process heat utilisation in industry
5. Solar water heating technologies

6.   Adoption of GTE technologies

7.   Retrofitting of automobiles to use CNG

8.   Implementation of clean technologies.

Enhancing life of an electrical lamp is of current interest to the producer as well as the user. The manufacturers can expect a good market for such products and the public are the true beneficiaries. Use of a compact fluorescent lamp, CFL as it is called , would cut down the power consumption by 70% in homes. In a similar manner, low-energy heat pumps, refrigerators and computers are to be widely marketted and used. Space heating of industrial and domestic facilities using natural gas saves extensive energy as compared to the electricity. It may also be emphasised in this connection that improvements in the domestic cookers designs would largely save energy. The use of natural gas in lieu of electricity by the coastal hotel industry in Western India has demonstrated the saving of energy. Waste heat/ heat generating substances arising from industrial process enable increase the productivity. Excessive use of solar water heaters introduce power saving. Where possible, hot steam or water coming from geothermal springs be used for minor purposes by the local population. Retrofitting of automobiles to adapt CNG in lieu of petrol be encouraged in order to reduce fuel bills. Finally, implementation of clean technologies, such as 'fuel cell' or hybrid-fuelled cars, buses be promoted by the respective governments with subsidies etc.

AUDIT PLANS

Management Audit

As in any environmental audit, the management of an Industry may appoint a team of experts to scrutinise all systems and evolve a *modus operandi* for plugging the loss of energy, or achieving saving of energy. The level of responsibility is fixed by the senior official namely the Managing Director. A senior official is given the responsibility to coordinate the audit team's work. Necessary help from the technical personnel and the shop floor level staff is assured to the audit team.

The requisite education, training and communication skills were imparted to all the staff who contribute to the energy audit. The methodology is laid out for rapid assessment and finally the management gets the recommendations to be implemented immediately eventhough some financial expenditure is involved. The policies laid down by the management shall thus be progressive in order to achieve the objective.

Some examples of this type are: 1. The management of a transport company decides to run the buses with 50% occupancy during the thin hours by suitably adjusting the timings. 2. The DTV of a city insists on a minimum occupancy of 2 in each car that runs through a fast lane of the road. 3. A decision to allow a 5-day in a week work programme will extensively save electric power and fuel.

Material Audit

Saving of raw material, or intermediates shall be noted. The use of natural gas in lieu of electricity enables anyone or any organisation to save 'one step process fuel', as electricity production always involve burning oil or gas. Any energy substances saved in processes e.g., sulphur, metals, hydrogen are to be audited. Even 'carbon dioxide generation substances', if in large quantity are to be accounted as it is directly related to energy. Pollution reducing schemes automatically account for saving of energy in the processes. Use of catalytic converters increase the efficiency of fuel burning and so to be considered for audit.

Waste heat emanating from the systems that use pyrophoric or nuclear materials can also be successfully utilised to set different thermal reactions going in other systems. Thus a variety of useful products can be obtained as bi-products.

METHODOLOGY

The first step under this scheme is to record the sources of energy which operate in the establishment. This may be electricity or heat generated from oil, coal or natural gas or biomass. The estimate of solid, liquid and gas mass has to be recorded. The stock position at the company has to be noted. The tariff paid for the fuel also to be noted.

The manner in which the energy is used in the company e.g., for building heating or refrigeration, the duration, insulation technologies adopted, the efficiency of systems, whether automatic or manual, their wear and tear etc., has to be observed and reviewed. If the energy use pertains to a 'storage tank' the dimensions, area, ducting, insulation and operating temperatures are to be mentioned. If the work involves a 'process' such as boiler operation, furnace use, condensation, distillation, or engineering jobs as lifting, digging, earth moving, or melting, it shall be recorded together with a rough estimate of energy consumption and duration of the process. An energy flow diagram is quite useful for assessing the requirements.

Schemes for Energy Conservation: Depending on the urgency, the schemes are categorised into:

1. ***Short term:*** that involves rather low capital expenditure and bringing rapid results. The jobs involve cleaning and maintenance of oil burners, heat exchangers, furnaces etc. Good house-keeping such as tight doors and windows, insulations render preservation of heat in the systems. Leakages are to be plugged wherever steam is let in. Electric power also has to be conserved and monitored carefully in the short term schemes.

2. ***Medium term :*** schemes are those where some cost effective modifications for the equipment and machinary is effected. These are in the nature of improving the heating or steam delivering systems, process modification with alternative effective materials, modifying the process burners atomisation and re-designing the distribution of fuel.

3. ***Long term :*** replacement of heaters, pre-heaters with modern versions, use of high energy furnaces, innovative designs for distribution of energy, effective heat recovery processes.

The schemes under 1 and 2 above render a saving of energy to the extent of 5-10% and the scheme under 3 can achieve an energy saving between 10-15% in any industry.

Energy Index

This is a figure which is obtained from the equation

Energy Index = Energy use/Production output

This index is calculated for each type of energy employed in an industry over the production periods, weekly or monthly or annual. Summing all these types, the total energy index can be deduced.

For calculation purposes , the unit of all types of energy is brought to one form, either Joule or watt. The production output may be in tons.

Energy Costs

These are made on the basis of the individual costs of different types of energy systems used, namely oil, coal, natural gas, wood, electricity etc. The tariff for each one is different and a summation of all these gives the total cost of energy. In industry, it is a common practice to represent the energy consumption, energy cost on week basis in the form of either a 'pie' chart or 'pillar' chart for quick examination.

Energy Audit Report

The document 'Energy Audit Report' shall contain all the data which has been referred above pertaining to the said company/industry during a specified period. Recommendations with regard to improvement of conditions of the equipment, manpower skills, working environment and conservation of energy are also to be incorporated in the report.

Questions

1.　What are the types of environmental audits in practice? What are the advantages of audit?

2.　Explain how an environmental audit is conducted in a heavy chemical industry.

APPENDIX - 1

Renewable Energy Potential and Achievements in India.

Source	Unit	Potential, approx.	Achievements upto2000.
Small hydropower (upto 25MW)	M W	15,000	1,341
Wind power	M W	45,000	1,267
Solar water heaters	Million sq.m. collection area	140	0.55
Solar photovoltaics	MW/sq,km	20	47
Biomass power	M W	19,500	308
Biomass gasifiers	—	16,000	35
Biomass cogeneration	—	3,500	273
Biomass improved cookstoves	Million	120	33
Biogas plants	Million	12	3.1
Urban & Industrial waste based power	M W	1,700	15.20

Source: *TERI Energy Data Directory and Yearbook 2001/ 2002*

http://www.teriin.org/energy/tech.htm

SUGGESTED BOOKS & LIBRARY PAPERS

Books

MN.Sastry, *Energy resources and Environment*, Himalaya Publishing Co., Bombay, 1992.

RC Yadhav, *Energy Options and Environment*, Himalaya Publishing Co., Bombay, 1995.

H Sukhatme, *Solar Energy*, Tata McGraw Hill Co, Delhi, 1996.

Parulekar and Rao Anne L. Gaeperin, *Nuclear Energy-Nuclear Waste*, Chelsea House Publishers, New York, 1992.

Godfrey Boyle, *Renewable Energy*, Oxford University Press, Oxford, 1996.

Susanna Van Rose, *Earth*, DK Publishing Inc, New York, 1994.

Daniel B. Botkin and Edward A. Kellar,*Environmental Science- Earth as a Living Planet,* John Wiley & Sons, New York, 1998.

Robert I. Tilling, Volcanoes,http://pubs.usgs.gov/gip/volc/text.html

"Solar Living Sourcebook" , Ed. Doug Pratt & the Real Goods Staff, Chelsea Green Publishing Co., Vermount, 1999.

Martyn Bramwell, " Book of Planet Earth", Simon &Schuster, London, 1992.

Paul Gipe, "Wind Power for Home and Business", Chelsea Green Publishing Co, Vermont, 1993.

Mc Graw Hill Encyclopedia of Sci & Tech., 8th edition, **8**, "Hydrogen", Mc Graw Hill Book Co. New York, 1997; *ibid*, 7, " Fuel Cells", 1997.

Edward A. Keller, "Acid Rain", John Wiley & sons Inc. New York, p479,1995.

Shaw, RW, "Climate Changes", Scientific American, **257**, 96-103,1987; Weinberg, GJ and Williams, RW, Renewable Energy, *ibid*, **263**, 146-155, 1990.

Newsweek, USA, April 15,2002, "Beyond Oil- the future of energy", 32A-32R. **Contributors:** Fred Guterl; Adam Piore and Sandy Edry; Adam Piore; Stefan Theil; William Underhill; Erika Check.

MVRK Rao, "Environment in Submarines", DRDO Publication, 2002.

Joseph M. Norback, James W. Heffel, Thomas D. Drubin, Bessam Tabbara, John M. Bowden and Michelle C. Montano, "Hydrogen Fuel for Surface Transportation", Publ.by Soc. Automotive Engrs. Inc, Warrendale, PA, USA, 1992.

James C. White, "Global Energy Strategies", Plenum press, New York, 1993.

Van Nostrand's Scientific Encyclopedia, 8th edn. Ed. Douglas M. Considine, Van Nostrand Reinhold, New York, 1995, pp 706-728.

Mc Graw Hill Encyclopedia of Science & Technology, vol4, 9th edn, Mc Graw Hill Publishers, New York, 2002, pp 297-307.

Carl Heintze, " The Biosphere- Earth, Air, Fire and Water", Thomas Nelson Inc. Publishers, New york, 1977.

WR.Murphy and G.Ke Kay, " Energy Management", Chapter on Energy Auditing, Butterworths, Oxford, 2001.

INTERNET REPORTS

http://www.epa.gov/global warming/climate/index.html and other titles.

http://zebu.uoregon.edu/1998/es202/113 and 114.html

http://www.epa.gov/ozone/science/process.html

http://www.cpc.ncep.noaa.gov/products/stratosphere/sbuv2to/ozone_hole.html

http://zebu.uoregon.edu/1997/ph161/115.html ; /images/ozone 11.gif ozone reactions; /03 holesz.gif size; /ozdeclin.gif ozone deviation

http://library.thinkquest.org/17457/platetectonics/1.php; /2.php; /3.php;/4.php;/5.php

/volcanoes/types.composite.php

/erupt.php; /features.php and others

http://sci.org/impacts.html   coral reefs

http://www.crosswinds.net/~nqureshi/geothermalscan.html geothermal energy

http://volcano.und.nodak.edu/vwdocs/current_volcs/Barren_Isle.html

http://www.fe.doe.gov/international/indiover.html   overview

www.csis.org/saprog/sam14.html   south asia monitor

www.renewingindia.org/ethenv.html ; hyd.html; otec.html others

www.winrockindia.org/wedo3.htm.programs

www.autobahn.mb.ca/~het/energy.html all topics

www.epa.gov/oar/oarhome.html

www.nrdc.org/nrdc/dire/inxarch.html

http://haloedata.larc.nasa.gov/home.html ozone depletion; //spso2.gsfc.nasa.gov/spso-homepage.html

.noaa.gov/odp/— ;ncdc.noaa.gov/—-

http://www.ans.lneep.wise.edu/---;.rw.doe.gov/---waste managemnent; net.org/---industry; nrc.gov/---

.eren.doe.gov/---renewable energy; http://solstice.crest.org/---sustainable technology

http://daac.gsfc.nasa.gov/DAAC-DOCS/gdaac-home.html global warming

http://www.uic.com.au/ne6.htm chapters1,3,6

/peac. htm; ral.htm; nfc.htm; ueg.htm; wast.htm;whyu.htm; uicchem.htm

http://acre.murdoch.edu.au/ago/biomass/biomass.html; www.beral.org; org/about.html a renewable source.

www.epa.gov/docs/crb/apb/ biomass.htm biomass utilisation

www.ott.doe.gov/biofuels/what_are.html

//id.inel.gov/geothermal/articles/reed/index.html geothermal energy papers

//www.eren.doe.gov/geothermal/geobasics.html;/ geopowerplants.html ; /geodirectuse.html; /heatpumpsover.html; / ghp_buildings.html

http://waterpower.hypermart.net/index.html ; /wave.html; /otec.html; /waterwheels.html others

//news.bbc.co.uk/low/english/sci/tech/newsid_1032000/1032148.stm wave power station

//www.wavegen.co.uk/elictricl.htm; /product.htm; /tech.html

//www.eia.doe.gov/emeu/iea/table81.html oil, gas table for wworld countries

//www.fuelcells.org/what is.htm ; /fctypes.htm; fcapps.htm; fcbenifits; fuelcells.org/gallery.htm

http://americanhistory.si.edu/csr/fuelcells/alk/alkmain.htm; /phos/ pafcmain.htm; mc/mcfcmain.htm; so/sofcmain.htm; pem/ pemmain.htm; /future/furmain.htm

http://geoheat.oit.efdu/whatgeo.htm

http://www.geyserstudy.org/ov_quick_guide.htm;.org/ west_thumb.htm;

http://www.yellowstone.net/russfinley/yellow/geysers.htm

http://www.hsssi.com/Applications/Echem/Background/ waterelec.html ; /Echem/Hydrogen; /Oxygen

http://www.akeena.net/content/what_size_systems.htm

/akeena.net; /Backup_power_systems_ACGC.htm